Studies in Computational Intelligence

Volume 695

Series editor

Janusz Kacprzyk, Polish Academy of Sciences, Warsaw, Poland
e-mail: kacprzyk@ibspan.waw.pl

About this Series

The series "Studies in Computational Intelligence" (SCI) publishes new developments and advances in the various areas of computational intelligence—quickly and with a high quality. The intent is to cover the theory, applications, and design methods of computational intelligence, as embedded in the fields of engineering, computer science, physics and life sciences, as well as the methodologies behind them. The series contains monographs, lecture notes and edited volumes in computational intelligence spanning the areas of neural networks, connectionist systems, genetic algorithms, evolutionary computation, artificial intelligence, cellular automata, self-organizing systems, soft computing, fuzzy systems, and hybrid intelligent systems. Of particular value to both the contributors and the readership are the short publication timeframe and the worldwide distribution, which enable both wide and rapid dissemination of research output.

More information about this series at http://www.springer.com/series/7092

Roger Lee
Editor

Applied Computing and Information Technology

 Springer

Editor
Roger Lee
Software Engineering and Information
 Technology Institute
Central Michigan University
Mount Pleasant, MI
USA

ISSN 1860-949X ISSN 1860-9503 (electronic)
Studies in Computational Intelligence
ISBN 978-3-319-84659-0 ISBN 978-3-319-51472-7 (eBook)
DOI 10.1007/978-3-319-51472-7

Printed on acid-free paper

This Springer imprint is published by Springer Nature
The registered company is Springer International Publishing AG
The registered company address is: Gewerbestrasse 11, 6330 Cham, Switzerland

Foreword

The purpose of the 4th International Conference on Applied Computing and Information Technology (ACIT 2016) held during December 12–14, 2016 in Las Vegas, USA was to bring together researchers, scientists, engineers, industry practitioners, and students to discuss, encourage, and exchange new ideas, research results, and experiences on all aspects of Applied Computers and Information Technology, and to discuss the practical challenges encountered along the way and the solutions adopted to solve them. The conference organizers have selected the best 13 papers from those papers accepted for presentation at the conference in order to publish them in this volume. The papers were chosen based on review scores submitted by members of the program committee and underwent further rigorous rounds of review.

In chapter "Business Sustainability Conceptualization", Arunasalam Sambhan-than, Vidyasagar Potdar, and Elizabeth Chang document research that has been undertaken to define the term business sustainability by analyzing multiple defi-nitions proposed in the existing literature. The main conceptual contribution of this paper is its holistic presentation of the concept of business sustainability with contextual relevance to the ICT sector.

In chapter "A Replicated Study on Relationship Between Code Quality and Method Comments", Yuto Miyake, Sousuke Amasaki, Hirohisa Aman, and Tomoyuki Yokogawa clarify the relationship between comments other than SATD (self-admitted technical debt) comments and code quality. This study demonstrates a need for further analysis on the contents of comments and its relation to code quality.

In chapter "A Predictive Model for Standardized Test Performance in Michigan Schools", William Sullivan, Joseph Marr, and Gongzhu Hu build a predictive model as an early warning system for schools that may fall below the state average in building level average proficiency in the Michigan Educational Assessment Program (MEAP). They utilize data mining techniques to develop various decision tree models and logistic regression models, and find that the decision tree model with entropy impurity measure accurately predicts school performance.

In chapter "A Development Technique for Mobile Applications Program", Byeondo Kang, Boram Song, Seungwon Yang, and Jonseok Lee present a development technique, AppSpec, for mobile applications program. They applied AppSpec to developing a mobile application program, and then presented the products of diagrams as the result of performing development phases of AppSpec.

In chapter "Interactive Mobile Applications Development Using Adapting Component Model", Haeng-Kon Kim and Roger Y. Lee discuss some of the problems of the current mobile based human management applications and show how the introduction of adaptive component based development (CBD) model provides flexible and extensible solutions to mobile applications.

In chapter "Development of Guiding Walking Support Device for Visually Impaired People with the GPS", Tsubasa Sugimoto, Shota Nakashima, and Yuhki Kitazono conduct development and experiment of the walking support device that enable them to go out with ease by being equipped with sensors to detect obstacles and omni-wheels and navigate them in this study.

In chapter "User Evaluation Prediction Models Based on Conjoint Analysis and Neural Networks for Interactive Evolutionary Computation", Ryuya Akase and Yoshihiro Okada develop user evaluation prediction models based on conjoint analysis and neural networks for interactive evolutionary computation (IEC) implemented by interactive genetic algorithm and interactive differential evolution.

In chapter "Emotional Video Scene Retrieval Using Multilayer Convolutional Network", Hiroki Nomiya, Shota Sakaue, Mitsuaki Maeda, and Teruhisa Hochin propose a scene retrieval method based on facial expression recognition (FER). The effectiveness of the proposed method is evaluated through an experiment to retrieve emotional scenes from a lifelog video database.

In chapter "Proactive Approach for the Prevention of DDoS Attacks in Cloud Computing Environments", Badr Alshehry and William Allen propose a highly inventive multilayer system for protection against DDoS in the cloud that utilizes threat intelligence techniques and a proactive approach to detect traffic behavior anomalies. A series of experiments were performed and the results demonstrate that this multilayer approach can detect and mitigate DDoS attacks from a variety of known and unknown sources.

In chapter "Practical Uses of Memory Storage Extension", Shuichi Oikawa discusses the several practical uses of a memory storage extension (MSX) in cloud computing and also in the fundamental operating system architecture.

In chapter "How to Build a High Quality Mobile Applications Based on Improved Process", Haeng-Kon Kim and Roger Y. Lee identify defects to produce reliable mobile software and analyze the relationship among different defects among mobile applications. Another goal of this paper is to design a defect trigger based on the findings.

In chapter "A New Hybrid Discrete Firefly Algorithm for Solving the Traveling Salesman Problem", Abdulqader M. Mohsen, and Wedad Al-Sorori propose a new hybrid variant of discrete firefly algorithm, called HDFA, to solve traveling salesman problem (TSP). In the proposed improvement, the balance between intensification and diversification is achieved by utilizing the local search procedures, 2-opt and 3-opt, to improve searching performance and speed up the convergence.

In chapter "Empowering MOOCs Through Course Certifying Agency Framework", Yeong-tae Song, Yuanqiong Wang, and Yongik Yoon propose a MOOC course certifying agency framework, which merges learners' profiles from various MOOC providers so consolidated profiles are available in one place. The result is the list of learners who match or almost match a given job description.

It is our sincere hope that this volume provides stimulation and inspiration, and that it will be used as a foundation for works to come.

December 2016   Takaaki Goto
Ryutsu Keizai University
Ryugasaki, Japan

Contents

Contributors

Ryuya Akase Graduate School of Information Science and Electrical Engineering, Kyushu University, Nishi-ku, Fukuoka, Japan

Wedad Al-Sorori University of Science and Technology, Sana'a, Sana'a, Yemen

William Allen School of Computing, Florida Institute of Technology, Melbourne, FL, USA

Badr Alshehry School of Computing, Florida Institute of Technology, Melbourne, FL, USA

Hirohisa Aman Center for Information Technology, Ehime University, Matsuyama, Japan

Sousuke Amasaki Faculty of Computer Science and Systems Engineering, Okayama Prefectural University, Soja, Japan

Elizabeth Chang Australian Defense Force Academy, University of New South Wales, Canberra, Australia

Teruhisa Hochin Department of Information Science, Kyoto Institute of Technology, Kyoto, Japan

Gongzhu Hu Department of Computer Science, Central Michigan University, Mount Pleasant, USA

Byeondo Kang Department of Computer and Information Technology, Daegu University, Gyeongsan, Republic of Korea

Haeng-Kon Kim School of Information Technology, Catholic University of Daegu, Daegu, Korea; Department of Computer Engineering, Catholic University of Daegu, Daegu, Korea

Yuhki Kitazono Department of Creative Engineering, National Institute of Technology, Kitakyushu College, Kitakyushu, Japan

Jonseok Lee Department of Computer Engineering, Woosuk University, Wanju County, Republic of Korea

Roger Y. Lee Department of Computer Science, Central Michigan University, Mount Pleasant, USA

Mitsuaki Maeda Department of Information Science, Kyoto Institute of Technology, Kyoto, Japan

Joseph Marr Department of Computer Science, Central Michigan University, Mount Pleasant, USA

Yuto Miyake Faculty of Computer Science and Systems Engineering, Okayama Prefectural University, Soja, Japan

Abdulqader M. Mohsen University of Science and Technology, Sana'a, Sana'a, Yemen

Shota Nakashima Graduate School of Science and Engineering, Yamaguchi University, Ube, Japan

Hiroki Nomiya Department of Information Science, Kyoto Institute of Technology, Kyoto, Japan

Shuichi Oikawa Faculty of Engineering, Information and Systems, University of Tsukuba, Tsukuba, Ibaraki, Japan

Yoshihiro Okada Innovation Center for Educational Resource, Kyushu University, Nishi-ku, Fukuoka, Japan

Vidyasagar Potdar School of Information Systems, Curtin Business School, Curtin University, Perth, Australia

Shota Sakaue Department of Information Science, Kyoto Institute of Technology, Kyoto, Japan

Arunasalam Sambhanthan School of Information Systems, Curtin Business School, Curtin University, Perth, Australia

Boram Song Department of Computer and Information Technology, Daegu University, Gyeongsan, Republic of Korea

Yeong-tae Song Department of Computer and Information Sciences, Towson University, Towson, USA

Tsubasa Sugimoto Department of Creative Engineering, National Institute of Technology, Kitakyushu College, Kitakyushu, Japan

William Sullivan Department of Computer Science, Central Michigan University, Mount Pleasant, USA

Yuanqiong Wang Department of Computer and Information Sciences, Towson University, Towson, USA

Seungwon Yang Department of Computer Engineering, Woosuk University, Wanju County, Republic of Korea

Tomoyuki Yokogawa Faculty of Computer Science and Systems Engineering, Okayama Prefectural University, Soja, Japan

Yongik Yoon Department of Multimedia Science, Sookmyung University, Seoul, Korea

Business Sustainability Conceptualization

Arunasalam Sambhanthan, Vidyasagar Potdar and Elizabeth Chang

Abstract Sustainability of ICT businesses is a timely topic in the context of the rapidly changing macro-environmental market conditions. This paper documents research that has been undertaken to define the term business sustainability by analyzing multiple definitions proposed in the existing literature. The definitions of business sustainability have undergone a process of abstractive decomposition which decomposes the definition into pieces and then constructs a new definition from the pool of decomposed phrases. Business sustainability is a multifaceted concept with different emerging perspectives that can be achieved through the concentrated efforts on organizational activities and processes. The outcomes of business sustainability could benefit both the organization as well as the environment. Well defined production methods, architectural strategy and the organizational commitment towards technological support are paramount for ensuring business sustainability. This study's significance is the contribution of a newer and comprehensive version of the definition the term business sustainability. The main conceptual contribution of this paper is its holistic presentation of the concept of business sustainability with contextual relevance to the ICT sector.

Keywords Business sustainability · Sustainability management · ICT businesses

A. Sambhanthan (✉) · V. Potdar
School of Information Systems, Curtin Business School, Curtin University, Perth, Australia
e-mail: asambhanthan@acm.org

V. Potdar
e-mail: Vidyasagar.Potdar@cbs.curtin.edu.au

E. Chang
Australian Defense Force Academy, University of New South Wales, Canberra, Australia
e-mail: elizabeth.chang@adfa.edu.au

© Springer International Publishing AG 2017
R. Lee (ed.), *Applied Computing and Information Technology*,
Studies in Computational Intelligence 695,
DOI 10.1007/978-3-319-51472-7_1

1

1   Introduction

Sustainability is a well-used and popular term, but its use lacks in clarity. There are a number of terms, such as sustainable development and triple bottom line, which are interchangeably used with the term 'sustainability' [1]. At the business level, sustainability is often equated with eco-efficiency [2]. Documented research exists on the review of sustainability terms and their definitions [3]. These authors give prominence to terms like cleaner production, pollution prevention, pollution control, and minimization of resource usage, eco-design and other similar terms. However, there is a considerable gap in the research literature in terms of defining business sustainability with a holistic view. Being a branch of the Sustainability discipline, business sustainability has quite a number of perspectives and angles that have been researched. Investigations of the business sustainability concept include consumer behavior, climate change, stakeholder management, innovation and strategy. In the context of consumer behavior and sustainability, there are studies on measuring the gaps between customers' expectations and their perceptions on green products [4]; investigation of factors influencing the sustainable consumption behaviors of rural residents [5]; the role of moral leadership for sustainable production and consumption [6]; survey and analysis of consumer behavior on waste mobile phone recycling [7] and the empirical investigation of green purchase behavior among the younger generation [8]. Definition of climate strategies for business [9] and the influence of stakeholders' power and corporate characteristics and social and environmental disclosure [10] are studies that document the relation of climate change and sustainability. The climate change related studies take more of an environmental sustainability angle when looking at sustainability. In terms of innovation, the link between eco-innovation and business performance [11] and sustainability oriented innovation in small and medium enterprises [12] are some notable studies that have been recently published. These studies look at the economic dimension of sustainability in the context of small and medium businesses. From a management perspective, there are studies on, for example, keeping track of corporate social responsibility as a business and management discipline with particular reference to Pakistan [13]; a study related to the strategic niche management of cleaner vehicle technologies from prototype to series production [14]; and the critical importance of strategic competencies for sustainable development [15]. The management related studies have been primarily focused on the strategic and sustainable development angles of the subject. However, there is still a lack of research in terms of integrating these disciplinary perspectives in sustainability related research.

Besides having a solid lexical meaning, the literature definitions of business sustainability are many. A recent survey of managers and business executives, published in the MIT Sloan Management Review, indicates that each company uses their own definition for the term sustainability [16]. The article further suggests that each company is focusing on specific perspectives in defining the term sustainability, such as only the environmental impact or else focusing on the economic, societal and personal implications. In fact, there could be a number of ways in which the term sustainability could be defined according to the context and perspective. However, it can be questioned whether the existing definitions of Corporate Sustainability could be substituted to refer to business sustainability. In fact, corporate sustainability and business sustainability are more like different expressions of the same phenomenon according to some critics. The critics could further argue that the term corporate sustainability is a widely used concept in contemporary management literature. However, the term corporate sustainability is observed to have a number of inherent difficulties in getting substituted for the term business sustainability. [17] Documented a number of observations regarding the literature related to the term corporate sustainability as listed below.

- The term 'Corporate Sustainability' is more widely used in specialized academic literature than in practitioner and top academic management literature
- A standardized definition of corporate sustainability does not exist
- Corporate sustainability has been conceptualized using different theoretical approaches
- A standardized method to measure corporate sustainability does not exist [17].

We anchor our argument for defining the term business sustainability in a more holistic and concise manner based on these observations.

1. We highlight the absence of the term 'corporate sustainability' across different types of management literature, which raises the need for a more focused effort in defining business sustainability with a better practitioner focus.
2. The absence of a standardized definition of corporate sustainability suggests the emerging need for defining business sustainability, which is a relatively newer term than corporate sustainability.
3. The term corporate sustainability has been conceptualized using different theoretical approaches which suggest the same could also be applied for business sustainability.
4. A standardized method to measure corporate sustainability does not currently exist in the literature, which shows that the concept of business sustainability has room to be developed.

Hence, we argue that a practitioner centered, multi-perspective, standard and holistic definition for business sustainability is a valuable contribution to the business sustainability research literature. Therefore, this article aims at crafting a contextual definition of business sustainability by analyzing multiple definitions from different perspectives. An abstractive decomposition of the term has been undertaken through the conceptual review of the literature, which eventually leads towards a holistic definition of business sustainability.

2 Business Sustainability—Definitional Paradox

Sustainability in business has a number of perspectives for which the definition has been constructed such as production, operation, supply chain and value. We have identified ten perspectives on the basis of a first level analysis of the business sustainability definitions (Table 1). In the context of this paper, the term perspective refers to the way in which we look at the sustainability phenomenon. In particular, the dominating aspect of the definition which signifies the overall theme of the definition has been referred to as perspective in this research. The definitions which are proposed by past researchers for the term "sustainability" have been anchored on these perspectives. Ideally, the dominating phrase of each definition has determined the selection of perspectives and the categorization of definitions into the relevant perspectives. We selected the term sustainability as the yardstick of evaluation instead of business sustainability for the following reasons:

- There are limited numbers of definitions available in the literature for the term 'Business Sustainability' [17].
- Business Sustainability is a type of sustainability which is appropriate for businesses.

Table 1 Ten perspectives of business sustainability

	Perspective	Rationale
1	Development	Any business will require ongoing research and development in order to sustain themselves in the rapidly increasing competitive market
2	Supply Chain	The supply chain of any business will need to be sustainable in order to leverage the maximum business outcome
3	Production	The production process of a business needs to be sustainable in order to harness the optimum productivity
4	Operation	The business operation of any firm needs to be sustainable in order to ensure maximum customer satisfaction
5	Stakeholder	The stakeholders of the business are the key players who decide the strategic direction of the business. This means that the stakeholders could influence the sustainability of business to an extent
6	Value	The value of the products needs to be ensured to have sustainable value generated in the long run
7	Organizational Structure	The organizational structure of any business needs to be sustainable in order to maintain maximum productivity
8	Business Strategy	The business strategy is paramount in defining the sustainability of its direction
9	Success Mechanism	The success mechanism needs to be sustainable in order to maintain a higher growth rate of the business
10	Management	The management of any business needs to be sustainable in order to make the overall functioning of the business move smoothly

Whilst defining the scope for definition filtering, as mentioned above, we have focused our search on original business sustainability definitions. In addition to this, we have selected ten perspectives for anchoring our evaluation which are depicted in Table 1. For example, [18] suggests "Business sustainability relies on the integration of six dimensions of sustainability (Social, Ecological, Economic, Spatial, Institutional-political and cultural) into mainstream decision making and core operational processes as the articulated systems". Based on this definition, operation has been identified as one of the key perspectives. Arguably, the dominating idea in this definition is operation. Table 1 depicts the ten perspectives identified in the literature and justifies the rationale for each selection.

Individual definitions abstracted from each perspective have been analyzed in order to gain more insight into the definitions concerned. Table 2 depicts the twelve definitions of business sustainability from the past literature. A textual decomposition of each definition has been undertaken to elicit key conceptual elements from the definition which resulted in seventy-seven distinct conceptual elements (Table 2). Firstly, we have abstracted business sustainability definitions from a number of downloaded articles. Then we abstracted the definitions and did a textual decomposition of the definitions. We repeated this process until the definitions came to a stage where there was no new concept element to be derived. Finally, we numbered the concept pieces and organized them into a table (refer to Table 2). There were no specific criteria used for choosing the definitions, but we ensured that the maximum possible abstraction was completed by making sure that there were no concept pieces remaining. It is possible that relatively little data could be enough to draw valid conclusions in a qualitative research. Arguably, each definition considered in this evaluation could be considered as a standalone case sample for definitions used widely in the relevant literature. The depth of the analysis and the pattern matching possibility of themes and concept clusters is what matters for determining the number of samples in a case study based on qualitative research. The authors have made an attempt to deeply analyses the definitions though an abstractive decomposition process performed at both text and phrase levels. Therefore, the method adapted for this concept construction research could be argued to be more scientific than random.

3 Evaluation of Definitions

This section critically evaluates each perspective mentioned in the table below and the respective phrases that enable an intelligent selection of themes and sub-themes to be included in the new definition. The discussion section gives a definite justification for the new additions, and deletions from each definition concerned. In fact, certain definitional elements do not really serve the purpose of defining the term business sustainability and hence are excluded. Arguably, the context in which the definition is applied is more critical than the wording [1].

Table 2 A textual decomposition of the definitions of business sustainability abstracted from the literature

Perspective	Definition	Conceptual elements
Development	"The scope of the concept of Sustainability relies on the environment, society, economy, organization and the people within the organization", [19]	Environment[1] Society[2] Economy[3] Organization[4] People[5]
Supply Chain	"Sustainability is the total effort of a company (including its demand and supply chain networks) to reduce the impact on the Earth's life- and eco-systems", [20]	Company's effort[6] Demand[7] Supply chain network[8] Reducing impact on earth's life[9] Reducing impact on ecosystem[10]
Production	"Sustainable production is the creation of goods and services using processes and systems, that are non-polluting, conserving of energy and natural resources, economically viable, safe and healthful for workers, communities and consumers, and social and creatively rewarding for all working people", [21]	Production[11] Creation of goods and services [12] Processes[13] Systems [14] Non-polluting[15] Energy conservation[16] Natural resource conservation[17] Economic viability[18] Community safety[19] Consumer safety[20] Community health[21] Consumer health[22] Worker safety[23] Worker health[24] Socially rewarding for working people[25] Creatively rewarding for working people[26]
Operation	"Business sustainability relies on the integration of six dimensions of sustainability (Social, Ecological, Economic, Spatial, Institutional-political and cultural) into mainstream decision making and core operational processes as the articulated systems", [18]	Social[27] Ecological[28] Economic[29] Spatial[30] Institutional-political[31] Cultural[32] Operational processes[33] Decision making[34]
Stakeholder	"Meeting the needs of a firm's direct and indirect stakeholders, without compromising its ability to meet the needs of future stakeholders as well", [2] "Business sustainability seeks to create long-term shareholder value by embracing the	Meeting a firm's needs[35] Direct stakeholders[36] Indirect stakeholders[37] Future stakeholders [38]

(continued)

Table 2 (continued)

Perspective	Definition	Conceptual elements
	opportunities and managing the risks that result from an organization's economic, environmental, and social responsibilities", [23]	Create long-term shareholder value[39] Embracing opportunities[40] Managing risks[41] Economic responsibilities[42] Environmental responsibilities[43] Social responsibilities[44]
Value	"Sustainability is a business approach that seeks to create long-term value for stakeholders, by embracing opportunities and managing risks associated with economic, environmental and social developments", [24]	Business approach[45] Create long-term value for stakeholders[46] Embracing opportunities[47] Managing risks[48] Economic risks[49] Environmental risks[50] Social risks[51] Risks in social development[52]
Organizational Structure	"A sustainable organization, in addition to focusing on economic performance, actively supports the ecological viability of the planet and its species, contributes to equitable and democratic practices, and social justice", [25]	Economic performance[53] Ecological viability of the planet[54] Ecological viability of the species[55] Equitable and democratic practices[56] Social justice[57]
Business Strategy	"Business strategies that are intended to add social and/or environmental value to external stakeholders while increasing value to shareholders", [26]	Business strategies[58] Social value[59] Environmental value[60] External stakeholders[61] Increasing value to stakeholders[62]
Success Mechanism	"Sustainability is something which leads to medium-and long-term success based on concentrating on core competencies, on perusing right target groups, on finding suitable revenue models, and on designing appropriate products", [27]	Medium-and long-term success[63] Concentrating on core competencies[64] Perusing right target groups[65] Finding suitable revenue models[66]

(continued)

Table 2 (continued)

Perspective	Definition	Conceptual elements
		Designing appropriate products[67]
Management	"Successful Management of Business Sustainability therefore has to involve co-ordination of product design, manufacturing, delivery, distribution and disposal throughout the product life cycle", [28]	Management of business sustainability[68] Coordination of product design[69] Coordination of manufacturing[70] Coordination of delivery[71] Coordination of distribution[72] Coordination of disposal[73] Product life cycle[74]

Development is a critical factor for success in any organization. [19] covers five main clusters of the term sustainability; the environment, society, economy, organization and people. These elements are the most common in nature, and they overlap with some of the later definitions. Thus, there needs to be no additions required in this section. It could be questioned whether this definition is solely of sustainability or business sustainability. However, the usage of the term organization in this definition could increase its significance in this context.

Supply Chain covers the entire business process from production to service delivery. [20] define sustainability from a supply chain perspective, covering a broad spectrum of activities that fall under the entire supply chain framework of a business. The definition has *five* conceptual elements, but lacks emphasis on the *Demand and Supply Balance*[75] in terms of the production, delivery and consumption by the end user being paramount. Therefore, an inherent need for maintaining a proper balance between these activities is warranted, which would make this effort sustainable in terms of maintaining a smooth demand-supply network. The demand and supply balance is something that emerged as part of our own reflection which could add value to the pool of concept themes emerging as part of the new definition building.

Production is the core of any business endeavor. [21] adapted the LCSP definition of sustainability from a production perspective as comprising *sixteen* conceptual elements which are connected to the production that have substantial coverage of the people aspect of the production process, but has overlooked the essence of the other aspects such as *Infrastructure*[76] and *Methods*[77]. Arguably, the infrastructure and production methods adapted for each product development could add value to the overall sustainability of a production process.

Operation plays an important role in business sustainability. [18] define sustainability from an operation perspective which covers *eight* conceptual elements.

The definition includes the institutional-political dimension which could be looked at separately at the macro and micro level. This gets operationalized into the inter-institutional politics (at the macro level) of competitive forces and market penetration rivals and intra-institutional politics (at the micro level) within the organization for limited/shared resources. However, the definition fails to incorporate the *Technological*[78] factors, which are very critical for ensuring sustainability in an operation environment. In addition to this, architecture could also be included into the definition of sustainability. In fact, the architecture and the architectural strategy are the two main themes which make an impact on the business' sustainability. Arguably, one of the main contributions of the doctoral work of [22] is that the engineering design of each product evolves over time as the complexity evolves. [22] applies the theory of evolution to the engineering design process that defends the claim of design evolution in the context of an engineered production process. The research further proposes the term 'architectural strategy'. In the context of this research, the *Architecture*[79] could be made another additional element to the concept of sustainability due to the influence it has on the long-term sustainability of any business.

Stakeholders are the key people who make a difference in any business. [2] define sustainability from a stakeholder perspective with *four* conceptual elements. The concept themes which get emphasized in this definition are stakeholder and needs, with the needs the most prominent. Needs drives business functioning in any industry as it is the starting point of product design. Consumer needs create mind images of the product and this then gets formulated into a logical design which emerges into the physical design of the product. The definition also addresses another concept theme of 'stakeholder'. This theme has been clustered into three categories in terms of time and involvement. In fact, the stakeholders are categorized into three on the basis of the time and involvement (i.e., time—future stakeholders; involvement—direct stakeholders and indirect stakcholders). A possible addition to this could be the inclusion of other time-based classifications (i.e., past and present stakeholders). Arguably, the *Past Stakeholders*[80] and *Present Stakeholders*[81] could contribute towards overall sustainability through effective and constructive feedback. Another aspect of involvement-based classification (i.e., internal and external) could add another two elements to the theme. Thus, *Internal Stakeholders*[82] and *External Stakeholders*[83] are two other additions to the list. In contrast, the definition constructed by [23] includes factors such as risks, responsibilities, opportunities and value which are covered, to an extent, in the other definitions.

Value is critical in ensuring business sustainability of any kind. [24] defines sustainability from a value perspective consisting of *eight* conceptual elements which could be clustered into the themes of risks, opportunities and value. The above definition overemphasizes risk, indicating the predictive capacity of the risk factor in determining the value of any sustainable business. In contrast, it is questionable whether risk is the only factor which contributes towards value sustainability. However, the other two conceptual elements which have a thematic focus on stakeholders and opportunities have been covered in some of the previous definitions. Therefore, the overemphasis on the risk factor will not have any considerable impact on this context.

Organizations are made of human beings, requiring the need for governance to be paramount. [25] define sustainability from an organizational perspective incorporating *five* conceptual elements. The definition covers two main themes, namely governance and environment. In fact, these two themes are essential concerns of a sustainable business. However, the governance theme could well be focusing on how well an organization could contribute towards the overall governance excellence of the country as a whole. In other words, effective governance of an organization could make a contribution towards the effective governance of the country at large. However, the environmental aspect is neglected in this context, which has been covered in one of the previous sections.

Strategy plays a vital role in business success. [26] defines sustainability from a strategic perspective with *five* conceptual elements. Value and Stakeholders are the main themes around which the definition is centered. These two themes have been covered in advance in some of the previous sections. On the other hand, Success is the end goal of any business. [27] define sustainability from the perspective of success with *five* conceptual elements. The main themes, product design and opportunities, have already been covered in a previous definition. In contrast, the overall management of the business has an impact on its sustainability. [28] define sustainability from a management perspective with *seven* conceptual elements. Coordination (a functional theme) is focused as the critical conceptual element in this definition.

4   A Holistic Definition of Business Sustainability

Business sustainability is a multifaceted term with three main themes making up the whole concept. The classification of conceptual elements has been done based on the following questions. The questions are used as the yardsticks for classifying these elements into three clusters which would lead towards the formation of the new definition. The following triangle shows the dimensions in which the themes are defined.

1. Why Business Sustainability?
2. How can Business Sustainability be achieved?
3. Where will it lead the Business?

The above questions will address the three dimensions of a clear definition. Firstly, the reason for businesses to be sustainable is covered in the first question. This will include the goal orientation aspect of a good definition. In fact, the themes deal with the underlying motivations of the organizations with regard to achieving sustainability (i.e., organizational requirements). Secondly, the way in which businesses could attain sustainability is covered in the second question. This will cover the underlying processes, systems, methodologies and architecture which enable sustainability in a business. Thirdly, the ultimate outcome of business sustainability is covered. In other words, the deliverables of a sustainable business are covered in this question. The following definition of Business sustainability is crafted from the abstractive decomposition process. It addresses the aforementioned three questions.

Business Sustainability could be achieved through efforts that are concentrated on;

1. core competencies,
2. effective production,
3. embracing opportunities,
4. exploring suitable and sustainable revenue models,
5. focusing on economic performance,
6. managing risks,
7. pursuing right customer groups,
8. supporting ecological viability in the existing infrastructure processes and systems,
9. resource conservation using technology,
10. growth in business,
11. improved health and safety,
12. reduction of environmental impact,
13. increased production value,
14. rewarding employees for their performance, and
15. Maintaining long-term support for organizational performance.

5 Key Terms

- **Core Competencies**: The total accumulation of skills, methods, know-how and capabilities a company possesses in order to produce its intended products or services. Examples include the skill level of the software engineers in a software business.
- **Effective Production**: The process of creating the core business outcome using the methods, technologies and infrastructure present in the company.
- **Embracing Opportunities**: The activity which aims at tapping into any possible opportunity in terms of being and becoming a sustainable business.
- **Locating a Sustainable Revenue Model**: The structure and approach through which the return on investments could be leveraged.
- **Managing Risks**: Management of risks involved in the process of getting the production process smooth.
- **Resource Conservation Using Technology**: Any efforts associated with the conservation of resources using technology as a tool for those efforts.

- **Growth in Business**: Any sorts of growth related to the overall business and its functioning that are in common.
- **Health and Safety**: The overall health and safety of the employees, stakeholders and other species in the environment in which the business is operating.
- **Environmental Impact**: The overall emissions, effect and impact on the environment by the business.
- **Production Value**: The value added and increased to the business and the environment by the production related activities.
- **Long-term support for Organizational Performance**: The continuous and ongoing support to the organizations' performance by the management and its stakeholders.

The above definition highlights three main dimensions of business sustainability. Firstly, the goal or stage orientation of the term could be observed in the definition. In fact, business sustainability could be argued as a business goal or stage for businesses to aspire to. Secondly, the process dimension is notable. The entire definition speaks about a process mechanism through which business sustainability could be achieved. Thirdly, business sustainability could be argued as an approach through which a strategic business transformation could be made. The above definition could be used as an enabling philosophy for the above transformation. Hence the definition includes three main dimensions of the term business sustainability as listed below.

- A business goal or stage
- A business process mechanism
- A strategic transformation approach

Hence, the above definition could be claimed to be a holistic definition of the term business sustainability, which incorporates a product view, process view and strategic view of the term. Thus, the comprehensiveness of the definition could be argued as being more concrete 'on the ground' from an overall business perspective. Applying this definition could leverage the optimum business outcomes in any business which aspires to become sustainable whilst maximizing their profit.

6 Managerial Implications and Industry Applications

The term business sustainability is well-used in the practitioner literature compared to corporate sustainability [17]. However, corporate sustainability has been highlighted as a strategic priority for businesses [29]. In addition to this, the KPMG report highlights *seven key organizational steps* needed to implement and benefit from corporate sustainability in a business. We draw our managerial implications from the new definition and use KPMG's seven steps for outlining a sustainability strategy for software businesses. These seven steps are to be synthesized with the core research outcome of this paper to provide context to practitioners and Chief Sustainability Officers managing large scale software businesses.

Managerial Implications.

- The managers could use this conceptualization for defining their business processes with a sustainability focus
- The managers could use the documented definition as a guidepost to improve their managerial effectiveness in terms of delivering value for their customers
- The new definition provides possible inputs towards conceptualizing and devising a sustainability strategy for organizations
- This could lend viability towards creating an organisation with a wide sustainability knowledge base to assist educate employees on sustainability matters
- The real world business examples given for the conceptual elements of the new definition could suggest more contextual and applied relevance to the managers and business policy makers

Application of KPMG's Steps.

1. *Using scenario planning to identify potential risks, explore new opportunities and focus on economic performance* [29]. Arguably, the new definition highlights managing risks and embracing opportunities as two elements of business sustainability. Risk management in software businesses could be undertaken through ensuring the commitment of top management, gaining optimum user commitment and clarifying the project requirements to all parties. Favourable government regulations, emerging market trends and macro environmental changes could be considered as possible opportunities for a business. The degree of monetary value added to a piece of software during the production process could be mentioned as an indicator for economic performance of a software business.

2. *Setting ambitious targets and leading by example to attain sustainability in a longer run* [29]. This could be achieved through focusing on core competencies such as the main skills, methods, know-how, capabilities and areas in which the business is excelling or where they have an advantage. [30] developed a pyramid model comprising skills and competencies of software professionals from a junior programmer to the IT director. Arguably, software firms make a practice of developing a skills inventory which consists of the core competencies and skills of software professionals in the company.

3. *Start measuring environmental inputs and productivity across the business* [29]. This can be achieved by measuring productivity and resource usage across all parts of the business from water usage to energy consumption. In a software business, this could be measured through looking at the overall resource usage in terms of peripherals, energy and other resources, such as human resources.

4. *Encouraging employee engagement both internally and externally to the business* [29]. This is argued to be improving the employee engagement in terms of gathering ideas, opinions and viewpoints at different levels. In the

software industry, this could be practiced by collaborative decision making from all levels of staff in terms of greening issues.

5. *Firms with experience of optimizing their own businesses have found this to be a rich source of expertise that can in turn help develop new products, services or innovations for clients* [29]. Effective knowledge management synced with decision support systems could facilitate this function in a software business.

6. *Exploration of other options which could result from sustainability, such as improved resource efficiency, cost reductions and risk mitigation* [29]. Software firms could do this by employing strategies for risk mitigation and resource efficiency.

7. *Brand enhancement through transparent reporting which would be shared with a range of stakeholders, from potential investors and shareholders to clients and business partners* [29]. Software businesses could make use of accounting and reporting practices which are acceptable to industry standards in maintaining transparency in their business activities.

7 Conclusions

This paper documents the application of a research methodology called abstractive decomposition to the task of developing a new definition for Business Sustainability, derived from ten different perspectives and various existing definitions. The resultant definition presented here addresses an aspect of research literature regarded by the author to represent a gap in verifiable literary or lexical evidence for shaping a holistic definition of business sustainability. The ten perspectives that underpin the existing definitions analyzed here are regarded as those most critical for a successful business operation. These include matters vital for business development such as, supply chain, production, operation, stakeholder, value, organization, strategy, success and management. Having covered all these aspects in the new definition presented here, it could be argued that this definition is dependable and deployable by any organization regardless of the industry in which it operates. Arguably, there is no specific contextual factor associated with the new definition, which may threaten its generalizability, but it focuses on defining the term in an abstract manner, so that it could be tailored to any industry sector as needed.

The description of key terms used in this new definition provides business policy makers and managers clarity of definition itself to ensure a more practical and applied focus. Examples were derived from real world management issues to make the terms contextualized to the software industry. Citations to past applied literature on these terms has been provided to enhance clarity and theoretical relevance of the examples used from the real world software industry. However, the research has not collected the viewpoints of industry precisians to analyze the practical limitations of trying to achieve sustainability through the terms outlined. Inclusion of real world samples from business data and validating these through the use of data gathered via interview with industry experts might validate the ideas presented here and reduce the impact of this limitation. In addition to this, the research reported herein documents a qualitative concept analysis of existing published materials to establish the interrelationship of Business sustainability constructs. Findings reported here provide a contribution to future development in this research field.

Whereas the new definition suits to any industry, the method used for deriving the definition has been devised to have minimal exclusions to any aspects of business sustainability. Having written a new definition of its kind through an abstractive decomposition process, the contribution of this paper is positioned to be a building block in the business sustainability research domain with a more strategic focus. However, upcoming research is warranted in terms of developing this into a comprehensive survey of business sustainability frameworks from different industries, which is expected to be developing a synergy between business sustainability theory and practice.

Appendix

See Tables 1 and 2.

References

1. S. McKenzie, "Social Sustainability-Towards Some Definitions," *Hawke Research Institute Working Paper Series*, 2004. [Online]. Available: https://atn.edu.au/Documents/EASS/HRI/working-papers/wp27.pdf. [Accessed: 28-Apr-2015].
2. T. Dyllick and K. Hockerts, "Beyond the business case for corporate sustainability," *Bus. Strateg. Environ.*, vol. 11, no. 2, pp. 130–141, 2002.
3. P. Glavič and R. Lukman, "Review of sustainability terms and their definitions," *J. Clean. Prod.*, vol. 15, no. 18, pp. 1875–1885, 2007.
4. S.-C. Tseng and S.-W. Hung, "A framework identifying the gaps between customers' expectations and their perceptions in green products," *J. Clean. Prod.*, vol. 59, pp. 174–184, Nov. 2013.
5. P. Wang, Q. Liu, and Y. Qi, "Factors influencing sustainable consumption behaviors: a survey of the rural residents in China," *J. Clean. Prod.*, vol. 63, pp. 152–165, Jan. 2014.
6. O. M. Vinkhuyzen and S. I. Karlsson-Vinkhuyzen, "The role of moral leadership for sustainable production and consumption," *J. Clean. Prod.*, vol. 63, pp. 102–113, Jan. 2014.
7. J. Yin, Y. Gao, and H. Xu, "Survey and analysis of consumers' behaviour of waste mobile phone recycling in China," *J. Clean. Prod.*, vol. 65, pp. 517–525, Feb. 2014.
8. M. Kanchanapibul, E. Lacka, X. Wang, and H. K. Chan, "An empirical investigation of green purchase behaviour among the young generation," *J. Clean. Prod.*, vol. 66, pp. 528–536, Mar. 2014.
9. H.-L. Pesonen and S. Horn, "Evaluating the climate SWOT as a tool for defining climate strategies for business," *J. Clean. Prod.*, vol. 64, pp. 562–571, Feb. 2014.

10. Y. Lu and I. Abeysekera, "Stakeholders' power, corporate characteristics, and social and environmental disclosure: evidence from China," *J. Clean. Prod.*, vol. 64, pp. 426–436, Feb. 2014.

11. C. C. J. Cheng, C. Yang, and C. Sheu, "The link between eco-innovation and business performance: a Taiwanese industry context," *J. Clean. Prod.*, vol. 64, pp. 81–90, Feb. 2014.

12. J. Klewitz and E. G. Hansen, "Sustainability-oriented innovation of SMEs: a systematic review," *J. Clean. Prod.*, vol. 65, pp. 57–75, Feb. 2014.

13. Z. A. Memon, Y.-M. Wei, M. G. Robson, and M. A. O. Khattak, "Keeping track of 'corporate social responsibility' as a business and management discipline: case of Pakistan," *J. Clean. Prod.*, vol. 74, pp. 27–34, Jul. 2014.

14. D. Sushandoyo and T. Magnusson, "Strategic niche management from a business perspective: taking cleaner vehicle technologies from prototype to series production," *J. Clean. Prod.*, vol. 74, pp. 17–26, Jul. 2014.

15. K. Mulder, "Strategic competencies, critically important for Sustainable Development," *J. Clean. Prod.*, vol. 78, pp. 243–248, Sep. 2014.

16. M. S. Hopkins, A. Townend, Z. Khayat, B. Balagopal, M. Reeves, and M. Berns, "The Business of Sustainability: What It Means To Managers Now," *MIT Sloan Manag. Rev.*, vol. 51, no. 1, pp. 20–26, 2009.

17. I. Montiel and J. Delgado-Ceballos, "Defining and Measuring Corporate Sustainability: Are We There Yet?," *Organ. Environ.*, vol. 27, no. 2, pp. 113–139, Apr. 2014.

18. C. H. Cagnin, D. Loveridge, and J. Butler, "An Information Architecture to Enable Business Sustainability," 1999.

19. Sutton Philip, "Sustainability: What does it mean?," *Green Innovations Inc*, 2000. [Online]. Available: http://www.green-innovations.asn.au/sustblty.htm#scope. [Accessed: 28-Apr-2015].

20. N. M. Høgevold and G. Svensson, "A business sustainability model: a European case study," *J. Bus. Ind. Mark.*, vol. 27, no. 2, pp. 142–151, 2012.

21. V. Veleva, "Indicators of sustainable production," *J. Clean. Prod.*, vol. 9, no. 5, pp. 447–452, 2001.

22. Woodard Charles Jason, "Architectural Strategy and Design Evolution in Complex Engineered Systems," 2006.

23. R. B. Pojasek, "A framework for business sustainability," *Environ. Qual. Manag.*, vol. 17, no. 2, pp. 81–88, 2007.

24. J. Galbreath, "Addressing sustainability: a strategy development framework," *Int. J. Sustain. Strateg. Manag.*, vol. 1, no. 3, pp. 303–319, 2009.

25. S. Benn and D. Dunphy, "Can democracy handle corporate sustainability? Constructing a path forward," Dec. 2014.

26. Reed Donald J., "Stalking the Elusive Business Case for Corporate Sustainability," 2001.

27. G. H. H. Breitner, *E-Learning*. Heidelberg: Physica-Verlag HD, 2005.

28. P. Hong, H. Kwon, and J. Jungbae Roh, "Implementation of strategic green orientation in supply chain," *Eur. J. Innov. Manag.*, vol. 12, no. 4, pp. 512–532, Oct. 2009.

29. KPMG, "Corporate sustainability: A progress report," Apr. 2011.

30. R. Colomo-Palacios, *Professional Advancements and Management Trends in the IT Sector*. IGI Global, 2012.

A Replicated Study on Relationship Between Code Quality and Method Comments

Yuto Miyake, Sousuke Amasaki, Hirohisa Aman and Tomoyuki Yokogawa

Abstract *Context*: Recent studies empirically revealed a relationship between source code comments and code quality. Some studies showed well-written source code comments could be a sign of problematic methods. Other studies also show that source code files with comments confessing a technical debt (called self-admitted technical debt, SATD) could be fixed more times. The former studies only considered the amount of comments, and their findings might be due to a specific type of comments, namely, SATD comments used in the latter studies. *Objective*: To clarify the relationship between comments other than SATD comments and code quality. *Method*: Replicate a part of the latter studies with such comments of methods on four OSS projects. *Results*: At both the file-level and the method-level, the presence of comments could be related to more code fixings even if the comments were not SATD comments. However, SATD comments were more effective to spot fix-prone files and methods than the non-SATD comments. *Conclusions*: Source code comments other than SATD comments could still be a sign of problematic code. This study demonstrates a need for further analysis on the contents of comments and its relation to code quality.

Keywords Source code comment · Software quality · Self-admitted technical debt

Y. Miyake · S. Amasaki (✉) · T. Yokogawa
Faculty of Computer Science and Systems Engineering,
Okayama Prefectural University, Soja 719-1197, Japan
e-mail: amasaki@cse.oka-pu.ac.jp

Y. Miyake
e-mail: cd28043k@cse.oka-pu.ac.jp

T. Yokogawa
e-mail: t-yokoga@cse.oka-pu.ac.jp

H. Aman
Center for Information Technology, Ehime University, Matsuyama 790-8577, Japan
e-mail: aman@ehime-u.ac.jp

© Springer International Publishing AG 2017
R. Lee (ed.), *Applied Computing and Information Technology*,
Studies in Computational Intelligence 695,
DOI 10.1007/978-3-319-51472-7_2

1 Introduction

The success of software development depends on many factors. The quality management of source code is one of the crucial activities to the success. Software developers dedicate much effort to retain the quality of source code from a spectrum of perspectives.

The comprehensibility of source code is essential for software developers working for a project where they maintain and evolve the source code cooperatively. For this purpose, the coding standards such as MISRA C [11] and Google Java Style [7] help to read source code and to prevent from faults though they provide no guide for code construction.

Comments in source code are useful documentation for developers and help their comprehension for the logic, architecture, and limitations of the code nearby. In fact, comments are widely known as effective entities [4, 14]. On the other hand, Fowler's book on refactoring [6] indicated that the presence of many comments is one of "code smells." Code smells are typical signs of poor design (poor quality parts of the code) to be refactored. Comments themselves have no direct effect on the design, but they often work as deodorants for smells. From a view of refactoring, comments thus play dual roles: well-written comments increase readability of programs, but they can also be signs of poor quality code.

Recent studies empirically revealed a relationship between source code comments and code quality. Comments themselves are harmless to code. Nevertheless, those studies both indicated some types of comments might be a useful sign for code quality improvement. Aman et al. focused on the relationship between the amount of code comments and code quality [1–3]. These analyses reported that well-written comments in a method body could be a sign of its fault-proneness. The relationship between the content of code comments and code quality was also studied [12, 13, 18]. These analyses focused on specific comments called *self-admitted technical debt* (SATD). SATD is an apparent quality risk confessed in code comments by developers. They first conducted manual examination for SATD [12, 13], and then revealed that SATD-related changes were difficult [18].

The studies [1–3] focused on the amount of comments and did not look into the contents of comments. The comments they analyzed contain the SATD comments, and it has been unknown that how much the SATD comments affected their results. A question thus arises whether the relationship between comments and code quality is viable for comments other than SATD comments. On the other side, the studies [12, 13, 18] were conducted on a bit rough-grained units, namely, source files. Comments usually help to understand a code fragment close to them, and a finer grained empirical study at the method level can show a detailed relationship between SATD and code quality.

In this study, we focused on a relationship between code quality and comments of methods. We extracted methods of classes and compared fix-proneness between the methods with comments other than SATD comments and those without any comments. We also conducted comparisons of fix-proneness between the methods with SATD comments and the other methods.

This paper is organized as follows. In Sect. 2, we present the design of our experiments. In Sect. 3, we show the results and their discussion. Related work is shown in Sect. 4. We conclude the paper in Sect. 5.

2 Methodology

2.1 Self-Admitted Technical Debt

Technical debts are introduced into source code by developers. Some technical debts are intentionally buried due to several reasons such as time pressure. A part of intentional technical debts are admitted in source code comments. Such technical debts are called *Self-Admitted Technical Debt* (SATD) [12]. An example of such comments is as follows:

```
if (implementation == null) {
    // No proper implementation
    // FIXME: We should log a warning, shouldn't we ?
    implementation = new AnyLanguage(language);
}
```

Here, a developer used a typical SATD keyword *FIXME* to share anxiety for logging option. In [18], such SATD keywords were used for specify SATDs.

2.2 Research Questions

This study aims at clarifying the relationship between comments other than SATD comments and code quality. SATD comments were rarely found in comments, and it was difficult to use its amount for experiments. Therefore, we decided to investigate a relationship between the presence of comments other than SATD comments and code quality. We used the fix-proneness to measure code quality as well as the studies [2, 18].

The studies [2, 18] conducted at different module levels: method and class. Comments usually describe a code fragment close to them, and the presence of a problematic fragment would also be signified by a comment near to it. This study thus focused on comments of methods but compared the fix-proneness at both the file level and the method level.

Table 1 OSS projects analyzed in the empirical work

Project	HEAD	KLOC	# of files	# of methods	# of commits
PMD	4e5e0187	90852	1048	6828	8698
SQuirreL SQL Client	7d867694	367732	2293	17802	7340
FreeMind	cc9cca40	113275	519	5631	1061
Hibernate ORM	ef46293c	453556	3326	28460	7083

We designed our empirical research on the following two research questions:

RQ1:   Are files containing comments other than SATD comments more fix-prone than files without any comments?

RQ2:   Are methods containing SATD comments more fix-prone than the other methods? Are methods containing comments other than SATD comments more fix-prone than methods without any comments?

2.3   Datasets

Our experiments required projects managed by a version control system because the fix-proneness required access to change histories of files. This study was conducted on four OSS projects: PMD,[1] SQuirreL SQL Client,[2] FreeMind,[3] and Hibernate ORM.[4] They all ranked in the top 50 popular Java products at SourceForge.net, are developed in Java, and their source files are maintained with Git. These projects were used in the past studies [2, 3]. Analyses on projects used in [18] are future work.

We selected Java files and focused on source code comments. The other files like XML were excluded from this study. Furthermore, our experiments compared the fix-proneness at two levels of source code, namely, file-level and method-level. Therefore, files not having any method were also excluded to keep the same files among experiments on the two levels.

Table 1 summarizes some statistics of project characteristics. The number of files counts the number of Java source code files which have at least one method. Table 1 shows a variety of projects in code size and the number of methods.

---

[1]http://pmd.sourceforge.net.
[2]http://squirrel-sql.sourceforge.net.
[3]http://freemind.sourceforge.net/wiki/index.php/Main_Page.
[4]http://hibernate.org/orm.

2.4 Target Comment Types

There are various types of comments described in source code [16]. This study includes experiments on the relationship between comments and code quality at the method-level, and we focused on comments related to methods: member comment and inline comment.

According to [16], member comments describe the functionality of method/field, being located either before or in the same line as the member definition. In our case, member comments are ones located ahead of a method. Member comments for fields were ignored. Member comments help developers understanding a usage or a specification of the method. Although Javadoc is a typical member comment, we analyzed member comments other than Javadoc. Because Javadoc rarely gives quality related descriptions and often ignored [3, 13].

Inline comments are located within a method body. Inline comments often describe code fragments nearby. Implementation decisions are also described by inline comments [16], they were candidates for a deodorant of code smells [3]. Sometimes code comments (commented-out code) were found within a method body. The code comments were irrelevant to our scope and ignored as well as the past studies.

2.5 Comments Extraction and Labeling

For each OSS project, we cloned the git repository. Table 1 shows HEAD sha of the projects. We then extracted methods and comments of them. For this purpose, we made a tool with Eclipse Java development tools (JDT).[5]

We first selected target files to be analyzed. The target files were source code written in Java. Files related to testing were excluded. The tool was applied to extract member and inline comments from the target files in each project. The tool also extracted method signatures and the range of method bodies. According to that information, the extracted comments were linked to the methods and classified into member comments, inline comments, and the others. Only the methods having member comments or inline comments were supplied for experiments.

For analyzing a relationship between comments and code quality, we have two options. The first one is to use the number of comment lines as well as Aman et al. [1–3]. The other is to use the presence of comments. This study focused on SATD comments and comments other than the SATD comments, and SATD studies [12, 13, 18] only considered the presence of SATD. Furthermore, SATD comments were rarely found in comments and it was difficult to use its amount for experiments. In this study, we thus decided to use the presence of SATD comments and the other comments for simplicity.

---

[5]http://www.eclipse.org/jdt.

Following [18], we first labeled the extracted methods as SATD or non-SATD with keyword matching. The keywords provided online by [12, 18] were used for the labeling. A method was labeled as SATD if any of the keywords appeared in member or inline comments of the method. Otherwise, the method was labeled as non-SATD. The non-SATD methods may contain comments other than SATD comments. A file was labeled as SATD if any of its methods were SATD.

We then labeled the non-SATD methods according to the presence of comments. A method was labeled as *commented* if the method has any comments. Note that no SATD keyword appeared in these comments. A file was commented if any of its methods were commented.

2.6 Fix-Proneness Identification

The fix-proneness were considered as a valuable measure for assessing code quality and maturity level [2, 18]. This study thus adopted the fix-proneness as code quality.

The definition of the fix-proneness defined as follows:

$$\frac{\# \text{ of fix commits}}{\# \text{ of commits}}$$

For each target file, all commits related to the file were obtained from a git repository. A commit was considered as a fix commit if any of bug-fixing-related keywords were found in a summary or a message of the commit. We used the following keywords shown in [18]: fixed issue #ID, bug ID, fix, defect, patch, crash, freeze, breaks, wrong, glitch, proper. All commits related to the target file were traversed, and the fix-proneness of the file was recorded.

The commits were then mapped to methods. We used `git diff` to identify lines added, modified, and deleted of a file. Matched the lines against the range of methods in the file, we identified methods changed by the commit. The fix-proneness of methods was obtained from the matching results.

3 Results

3.1 Are Files Containing Comments Other Than SATD Comments More Fix-Prone Than Files Without Any Comments?

Figure 1 shows boxplots of the percentage of defect fixing changes between files having and not having SATD methods for the four projects. This comparison works as a sanity check for [18] under a slightly different definition for SATD files and target

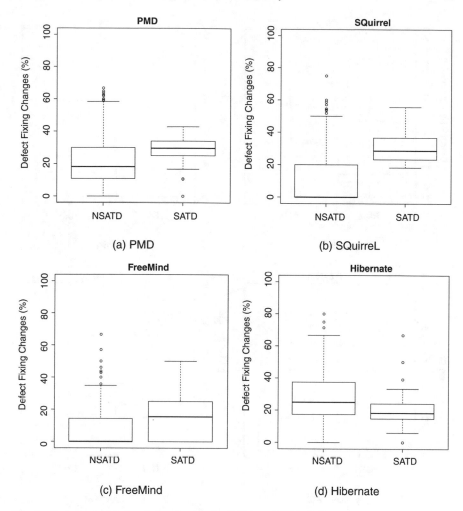

Fig. 1 Percentage of defect fixing changes for SATD and NSATD files

projects. All projects except for Hibernate (Fig. 1d) show that files having SATD methods have a higher percentage of defect fixing changes. This result was consistent to [18] somewhat where files having SATD comments were more fixed in four out of the five projects they used.

Figure 2 shows boxplots of the percentage of defect fixing changes between files having comments other than SATD comments and files having no comment for the four projects. As same as the previous results, all projects but Hibernate (Fig. 2d) show that files having comments other than SATD comments have a higher percentage of defect fixing changes.

Those differences were tested by Man-Whitney test [10] and quantified by the effect size with Cliff's delta [5] as well as [18]. All differences were statistically

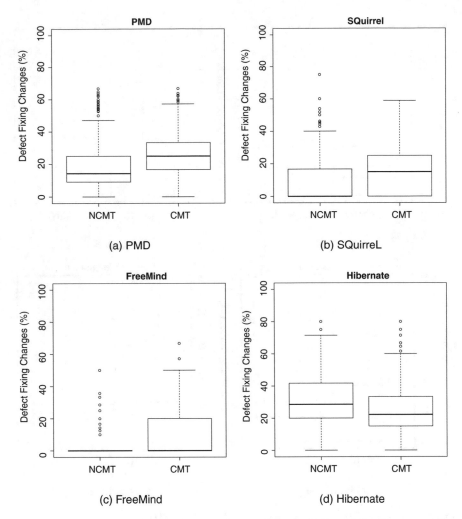

Fig. 2 Percentage of defect fixing changes for Commented (non-SATD) and non-Commented files

significant at $\alpha = 0.05$. Table 2 shows the effect-size for the four projects. Cliff's Delta ranges in the interval [-1, 1] and is considered negligible for $\|d\| < 0.147$, small for $0.147 \leq \|d\| < 0.33$, medium for $0.33 \leq \|d\| < 0.474$, and large for $\|d\| \geq 0.474$. The criteria indicate that the differences between SATD and non-SATD were medium or large for PMD, SQuirreL, and FreeMind. Only the difference on Hibernate was small. The differences between commented and non-commented files were all considered small.

Table 2 Cliff's Delta for SATD versus NSATD and Commented versus non-Commented files

Project	SATD versus NSATD	CMT versus NCMT
PMD	0.403	0.327
SQuirreL SQL Client	0.737	0.281
FreeMind	0.384	0.197
Hibernate ORM	−0.308	−0.192

These observations lead the following answer for RQ1:

- Where the presence of SATD comments increases defect fixing changes, the presence of comments other than SATD comments also increases defect fixing changes
- The difference of fix-proneness between commented and non-commented files was statistically significant but smaller than that between SATD and non-SATD files

3.2 Are Methods Containing SATD Comments More Fix-Prone Than the Other Methods? Are Methods Containing Comments Other Than SATD Comments More Fix-Prone Than Methods Without Any Comments?

Figure 3 shows boxplots of the percentage of defect fixing changes in SATD and non-SATD files for the four projects. This comparison shows a detailed relationship between SATD and code quality at the method level. Figure 3 shows that the fix-proneness was higher if methods had SATD comments for PMD and SQuirreL. This result was similar to the comparison at the file-level shown in Fig. 1. Contrastingly, the differences of the fix-proneness were less clear for FreeMind and Hibernate.

Figure 4 shows boxplots of the percentage of defect fixing changes in commented and non-commented methods for the four projects. As same as the results for SATD methods, commented files have a higher percentage of defect fixing changes for PMD and SQuirreL; the differences of the fix-proneness were less clear for Hibernate. For FreeMind, the difference could not be observed from Fig. 4c.

Those differences were tested by Man-Whitney test and quantified by the effect size with Cliff's delta. Regarding SATD methods, all differences except for Hibernate were statistically significant at $\alpha = 0.05$. Regarding commented methods, all differences including Hibernate were statistically significant at $\alpha = 0.05$.

Table 3 shows the effect-size for the four projects. The criteria indicate that the differences between SATD and NSATD were medium or large for PMD and SQuirreL. The difference for FreeMind was small, and that for Hibernate was negligible. The effect sizes at the method-level were smaller than those at the file-level.

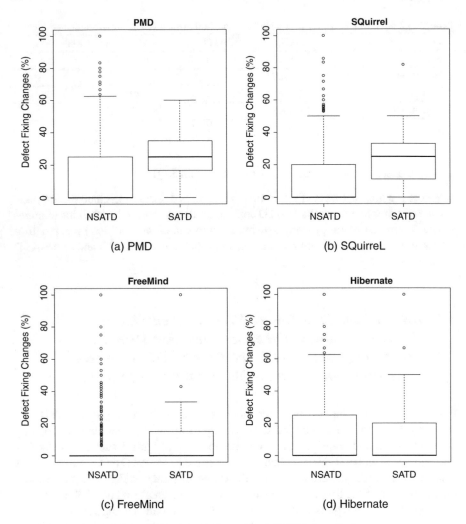

Fig. 3 Percentage of defect fixing changes for SATD and NSATD methods

The differences between commented and non-commented files were small for PMD and SQuirreL. The differences for FreeMind and Hibernate were negligible. These effect sizes were smaller than those on SATD methods. The effect sizes at the method-level were smaller than those at the file-level.

These observations lead the following answer for RQ2:

- Where the presence of SATD comments increases defect fixing changes at the file-level, defect fixing changes are also increased at the method-level
- Where the presence of SATD comments increases defect fixing changes at the method-level, the presence of comments other than SATD comments also increases defect fixing changes at the method-level

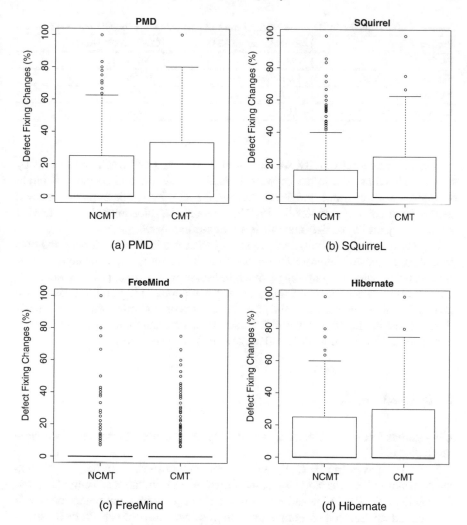

Fig. 4 Percentage of defect fixing changes for Commented and non-Commented methods

- The difference of fix-proneness between commented and non-commented methods was statistically significant but smaller than that between SATD and non-SATD methods

3.3 Threats to Validity

This study leaves some threats to validity as well as the past related studies.

Table 3 Cliff's Delta for SATD versus NSATD and Commented versus non-Commented methods

Project	SATD versus NSATD	CMT versus NCMT
PMD	0.381	0.273
SQuirreL SQL Client	0.514	0.156
FreeMind	0.170	0.096
Hibernate ORM	−0.03	−0.102

Regarding internal validity, we classified fix commits and SATD comments by keyword matching. While the keyword matching is simple and scalable, it might overlook the fix commits and SATD comments. However, this approach worked effectively in not a few studies [8, 15, 20]. Manual classification is labor-intensive for large projects in practice, and this is a realistic approach.

Regarding external validity, our target projects were limited in number and programming language. Therefore, the observations in this study might not hold in other projects using other programming language. However, the target projects were collected from various application areas, and most of popular programming languages are based on imperative manner and provide function or method constructs as a basis of development. Therefore, those differences are not critical for the generality of our contribution. Our future work includes experiments on such projects.

4   Related Work

Many studies focused on comments and its relation and effect to aspects of software development.

While comments can be freely written, its main purpose is to help developers comprehending the programs' logic, architecture, and limitations. Tenny [17] and Woodfield et al. [19] conducted experiments about program comprehensions, with students and experienced programmers as their subject, respectively. In their experiments, the subjects evaluated the ease of understanding some variations of a program which were different in modular design and in commenting manner (with or without comments.) Their experimental results showed that comments have a positive impact on program comprehension. The results also implies that the quality of comments also matters for program comprehension.

Steidl et al. [16] proposed a framework to evaluate the quality of comments. They evaluated the coherence between member comments and the name of the corresponding method. Their coherence evaluation presents whether the member comments provide useful information about the method. They discussed the comments written inside method bodies as well, and proposed to use short comments as indicators of parts to be refactored. Moreover, they described that the existence of long comments may infer a lack of external documents, so adding many comments was not recommended. Lawrie et al. [9] also studied the correspondence of comments and code by using a natural language processing technique, for an automated quality assessment.

Recent studies get focused on comments from the perspective of code quality. Aman et al. conducted some empirical studies of the relationships between the amount of comments and the fault-proneness [1, 3]. Their empirical results showed that the presence of comments written inside method bodies (they called "inline comments") is correlated with the fault-proneness in the programs. Therefore, well-written inline comments in a method body could be a sign of its fault-proneness as indicated as code smells by Fowler et al [6]. They also showed that the amount of comments contributed to explain the change proneness of programs [2].

The content of comments was also analyzed and related to the code quality. Potdar and Shihab [12] used a certain type of comments to specify technical debts. They called such commented technical debts as *self-admitted technical debt* (SATD). They manually examined source code comments of OSS projects and showed detailed statistics of SATD. Another study [13] regarding SATD classified SATD into five debts and showed what types of debts were common among their target projects. The relationship between SATD and code quality was examined in [18]. They used typical SATD keywords found in the previous studies to determine SATD files. Then, they related the presence and the introducing of SATD with defect inducing changes, fix inducing changes, and difficulty of code changes.

Our study heavily relies on the studies focused on the relationships between comments and code quality. While the studies by Aman et al. did not care about SATD, this study analyzed the quality of code having non-SATD comments. While Wehaibi et al. [18] considered the quality at the file-level, this study also focused on the quality at the method-level.

5 Conclusion

We replicated a study on the relationship between method comments and quality. The results show that the presence of comments could be a sign of problem even when the method comments did not contain SATD. We also revealed that the presence of SATD comments could also be a sign of problem even at the method-level.

Our future work includes further analyses on the relationship from another perspective. Defect inducing change was one of popular quality measures in literature including the past studies we replicated. The comparison of quality between pre- and post- comment inducing has also remained as future work. With deeper understanding on comments, we can expect a system suggesting risky or smelled code at writing comments.

Acknowledgements The authors would like to thank the anonymous reviewers for their thoughtful comments and helpful suggestions on the first version of this paper. This work was partially supported by JSPS KAKENHI Grant #16K00099.

References

1. Aman, H.: An Empirical Analysis on Fault-Proneness of Well-Commented Modules. In: Fourth International Workshop on Empirical Software Engineering in Practice (IWESEP). IEEE (2012)
2. Aman, H., Amasaki, S., Sasaki, T., Kawahara, M.: Empirical Analysis of Change-Proneness in Methods Having Local Variables with Long Names and Comments. In: ACM IEEE International Symposium on Empirical Software Engineering and Measurement (ESEM), pp. 1–4 (2015)
3. Aman, H., Amasaki, S., Sasaki, T., Kawahara, M.: Lines of Comments as a Noteworthy Metric for Analyzing Fault-Proneness in Methods. IEICE - Transactions on Information and Systems **E98.D**(12), 2218–2228 (2015)
4. Buse, R., Weimer, W.: Learning a Metric for Code Readability. IEEE Transactions on Software Engineering **36**(4), 546–558 (2010)
5. Cliff, N.: Ordinal Methods for Behavioral Data Analysis (1996)
6. Fowler, M., Beck, K., Brant, J., Opdyke, W., Roberts, D.: Refactoring. Improving the Design of Existing Code. Addison-Wesley (2012)
7. Google: Google java style. http://google-styleguide.googlecode.com/svn/trunk/javaguide.html
8. Kim, S., Whitehead Jr., E.J., Zhang, Y.: Classifying Software Changes: Clean or Buggy? IEEE Transactions on Software Engineering **34**(2), 181–196 (2008)
9. Lawrie, D.J., Feild, H., Binkley, D.: Leveraged Quality Assessment using Information Retrieval Techniques. In: the 14th IEEE International Conference on Program Comprehension (ICPC). IEEE (2006)
10. Mann, H.B., Whitney, D.R.: On a Test of Whether one of Two Random Variables is Stochastically Larger than the Other. The annals of mathematical statistics (1947)
11. MISRA: MISRA C. http://www.misra-c.com
12. Potdar, A., Shihab, E.: An Exploratory Study on Self-Admitted Technical Debt. In: IEEE International Conference on Software Maintenance and Evolution (ICSME), pp. 91–100. IEEE (2014)
13. da S. Maldonado, E., Shihab, E.: Detecting and quantifying different types of self-admitted technical Debt. In: IEEE 7th International Workshop on Managing Technical Debt (MTD), pp. 9–15. IEEE (2015)
14. Scanniello, G., Gravino, C., Risi, M., Tortora, G., Dodero, G.: Documenting Design-Pattern Instances: A Family of Experiments on Source-Code Comprehensibility. ACM Transactions on Software Engineering and Methodology **24**(3) (2015)
15. Śliwerski, J., Zimmermann, T., Zeller, A.: When do changes induce fixes? In: International Workshop on Mining Software Repositories (MSR), pp. 1–5. ACM (2005)
16. Steidl, D., Hummel, B., Juergens, E.: Quality analysis of source code comments. In: IEEE 21st International Conference on Program Comprehension (ICPC), pp. 83–92. IEEE (2013)
17. Tenny, T.: Program Readability: Procedures Versus Comments. IEEE Transactions on Software Engineering **14**(9), 1271–1279 (1988)
18. Wehaibi, S., Shihab, E., Guerrouj, L.: Examining the Impact of Self-Admitted Technical Debt on Software Quality. In: 2016 IEEE 23rd International Conference on Software Analysis, Evolution and Reengineering (SANER), pp. 179–188. IEEE (2016)
19. Woodfield, S.N., Dunsmore, H.E., Shen, V.Y.: The effect of modularization and comments on program comprehension. In: 5th international conference on Software engineering (ICSE), pp. 215–223. IEEE (1981)
20. Zimmermann, T., Premraj, R., Zeller, A.: Predicting Defects for Eclipse. In: International Workshop on Predictor Models in Software Engineering (PROMISE). IEEE (2007)

A Predictive Model for Standardized Test Performance in Michigan Schools

William Sullivan, Joseph Marr and Gongzhu Hu

Abstract Public school officials are charged with ensuring that students receive a strong fundamental education. One tool used to test school efficacy is the standardized test. In this paper, we build a predictive model as an early warning system for schools that may fall below the state average in building level average proficiency in the Michigan Educational Assessment Program (MEAP). We utilize data mining techniques to develop various decision tree models and logistic regression models, and found that the decision tree model with entropy impurity measure accurately predicts school performance.

Keywords Predictive modeling · Decision tree · Logistic regression · MEAP proficiency

1 Introduction

The ability of governing bodies to hold schools and school districts accountable to standardized test scores has been the subject of heated debate for many decades. Michigan adopted the Michigan Educational Assessment Program (MEAP) at the beginning of the 1969–1970 school year and administered the MEAP until 2013–2014. It was replaced with the Michigan Student Test of Educational Progress in 2015.

The MEAP test was taken every fall by public school students in grades 3–9. The test measured student proficiency in a number of different subjects. Due to budgetary constraints the only two subjects, reading and mathematics, in which every student

W. Sullivan · J. Marr · G. Hu (✉)
Department of Computer Science, Central Michigan University,
Mount Pleasant 48859, USA
e-mail: hu1g@cmich.edu

W. Sullivan
e-mail: sulli2wf@cmich.edu

J. Marr
e-mail: marr2ja@gmail.com

© Springer International Publishing AG 2017
R. Lee (ed.), *Applied Computing and Information Technology*,
Studies in Computational Intelligence 695,
DOI 10.1007/978-3-319-51472-7_3

is tested every school year. The MEAP test scores proficiency on a four point scale in which a score of 1 is highest and 4 is lowest. Students who score 1 or 2 are considered proficient and students who score 3 or 4 are considered not proficient.

Michigan holds schools accountable to the building level test scores of their students. The state tracks each schools percentage proficient and holds schools that fall below the state average accountable by either assigning them a "priority" ranking or by requiring "action plans" to help improve their test scores. In recent years state funding has been tied to student test scores as a way to increase levels of proficiency.

The aim of this research is to develop a predictive model that calculates building level proficiency scores of reading and mathematics and then predicts using different demographic factors the likelihood a school will achieve a proficiency percentage above the state average.

2 Related Work

The application of data mining to educational research is a relatively new endeavor. The field of Educational Data Mining (EDM), in general, is characterized by traditional data mining techniques and the inclusion of psychometric methodology [1, 2]. Much of the EDM literature has focused on improving student learning models or studying pedagogical support of learning software [3–5].

However, the application of classification techniques have also been applied to education research in the form of so called "Early Warning Systems". These systems have primarily been concerned with predicting high school drop outs. Bowers, Sprott, and Taff reviewed 110 proposed models and suggest standardized metrics for evaluation using latent class models [6]. Carl et al. developed a regression based "Early Warning System" which applies to a broader set of outcomes beyond dropping out of school [7]. Baradwaj and Pal focused on predicting student performance in higher education [8]. Knowles et al. take things a step further and utilize a number of ensemble methods including boosted tree and neural network frameworks [9].

In our study we built logit regression models and decision tree models to predict the students' performance in the MEAP tests.

3 Data

The data for this research comes from a number of different public sources:

(a) Michigan Student Data System (MSDS) [10]. The data include school types (traditional or charter), delivery method (traditional or virtual), etc.
(b) Michigan School Data [11]. The data set include teach/student ratio, Free and Reduced Lunch eligibility, and MEAP Proficiency scores by grade for all public schools in Michigan.

Table 1 Descriptive statistics of variables

Variable	Statistic				
	N	Mean	St. dev.	Min	Max
No. proficient	1,541	306	266	1	1,855
Percent proficient	1,541	0.510	0.185	0.023	0.989
Crime rate (CR)	1,541	0.065	0.053	0	1.117
Violent CR	1,541	0.006	0.006	0	0.025
Share of violent crime	1,541	0.079	0.050	0	0.269
% of population HS education or higher	1,541	87.251	6.041	67.300	99.400
Income	1,541	32,123	10,197	12,295	101,402
Free and reduced lunch (% eligible)	1,541	0.565	0.265	0.017	1.000
Free-lunch (% eligible)	1,541	0.510	0.267	0.012	1.000
Share of free-lunch	1,541	0.879	0.082	0.496	1.000
Teacher/student ratio	1,541	16.065	5.023	4.500	169.000
Jobless rate	1,541	7.520	1.818	4.700	12.700
State average (% proficient)	1,541	0.545	NA	NA	NA
Charter	218	NA	NA	0	1
Big 4	301	NA	NA	0	1

(c) Uniform Crime Report [12]. The report contains the total number of crimes along with details in over 350 cities in Michigan, including crime rate and share of violent crimes.

(d) American Community Survey (ACS) [13]. This survey provides information about jobless rate, education and income data for cities in Michigan.

From these data sets, we merged building level enrollment, demographic and income data with city level educational attainment and crime rate data to produce a sample of over 1,500 observations. This data set was used to predict the likelihood that a school achieves a proficiency percentage greater than the state average. The descriptive statistics of the variables are given in Table 1. In the following we shall describe these variables in more details.

3.1 School Proficiency Percentage

The Michigan Educational Assessment Program (MEAP) test was administered every fall to students in grades 3 through 9 from 1970–2014. The test evaluates a student's knowledge of the material covered the year before. The student is scored on a four-point scale where a score of 1 or 2 is deemed "proficient." The state uses

Table 2 Proficiency comparison of sample and population

Summary statistics	N	Mean
State percent proficiency	3,159	0.545
Sample percent proficiency	1,541	0.511

the percentage of students proficient as one of many ways to measure whether or not a school is effective. The percentage of proficiency of the population and sample is given in Table 2.

3.2 Crime Rate

The annual Uniform Crime Report (UCR) published by FBI provides detailed population and crime data on more than 350 cities in Michigan for 2014. For this research, it is apparent that when the data set is trimmed to include only schools located within cities in which UCR data is available the proficiency percentages of the sample are representative of the population as a whole. This results in a sample of just 1,541 observations.

Three crime rates (*CR*) are calculated:

$$CR = N_{crime}/N_{population}$$
$$Violent\ CR = N_{violent\ crime}/N_{population}$$
$$Share\ of\ Violent\ CR = N_{violent\ crime}/N_{crime}$$

where N_x is the number of x. These measures help separate the likely correlation between crime rate and other independent variables.

3.3 Education

The American Community Survey (ACS) is sent out to 3.5 million households each year and with this data the United States Census Bureau creates estimates of income and education. The estimates that were selected for the purposes of this research are estimates of the percentage of the population within a city that is 25 years or older and has obtained a high-school diploma or higher. The data from the ACS is then matched to each city with UCR data available resulting in a data set with 1,541 observations.

Table 3 Free and reduced lunch eligibility thresholds

FRL thresholds	Max income eligible (family of 4)	% in relation to poverty line
Free	$30,615	135
Reduced	$43,568	185

3.4 Income

The Michigan Department of Education (MDE) publishes a report detailing the number of students eligible for free and reduced lunch for each school every year. Table 3 shows the free and reduced lunch eligibility for 2014.

The percentage of students eligible for free and reduced lunch is used as a proxy for city level income data. This is appropriate because the student's eligibility is directly related to their family's income.

3.5 Class Size

Class size as measured by the student-to-teacher ratio for a school was calculated using the MDE's annual report on educator effectiveness and student count. It is traditionally considered an indicator of school quality. A number of schools are excluded because their primary method of delivery is online or they offer a significant number of online courses. Including these outliers would skew the data.

3.6 Jobless Rate

The number of people without a job has long been an economic indicator of great importance [14]. The ACS provides estimates for the jobless rate in every city in Michigan. The jobless rate of the city has been merged with all of the other demographic and building level information to help control for as many demographic variables as can be observed.

3.7 Charter Schools & Big 4 School Districts

A few dummy variables were created regarding the types of schools and their locations. One variable indicates whether or not a school is located in one of the "big 4" districts (Detroit, Lansing, Grand Rapids, Flint) in Michigan. Policy makers have long used these districts as a baseline in comparison to the others. The results and

Table 4 Sample comparison to charter schools and big-4

Summary statistics	N	Mean proficiency %
Charter public schools	218	37.2
Traditional public schools	1,249	53.4
Big 4 (traditional and charter)	213	29.7
Big 4 (charter only)	97	29.2

atmosphere of schools within these larger districts are considered more complex than the rest of the state. This may provide some interesting insight into the difference between schools that operate in an urban Michigan setting and schools that operate in a rural area. The other dummy variable was to identify charter schools. Table 4 shows the comparison of charter schools and "big 4."

3.8 Target Variable

State agencies often classify schools as underperforming or performing by whether or not performance is above or below the state average. The state average for school proficiency in our data is 54.5%. Thus, we generate a target variable called "Above Average" which takes on the value 1 for schools whose percent proficient is equal to or greater than 54.5% and 0 otherwise. Figure 1 provides a distribution of percentage proficient by schools. The distribution is mostly normal. Scores to the right of the mean (red line) are classified above average.

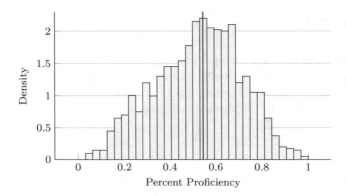

Fig. 1 Distribution—percentage proficient

Fig. 2 Work flow

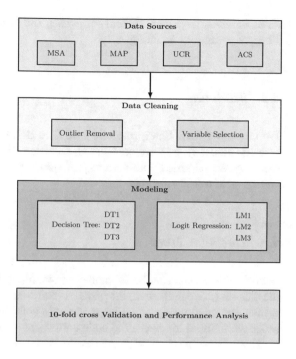

4 Methodology

The overall work flow of our analysis is shown in Fig. 2. We have described the data sources in the previous section, and shall discuss each of the remaining steps in this section.

4.1 Data Cleaning

Before analyzing the data, the data need to be trimmed and cleaned to include only schools operating within a city where all demographic information was available. The summary statistics show that the statewide proficiency percentage was within 3% of the samples average proficiency percentage. This means that the sample is representative of the population.

Prior to modeling, a filtering was applied to the data to remove observations with extreme outliers. As an example, one observation reported a Student/Teacher Ratio of 168, in contrast to the normal ratio approximately 17 for traditional classrooms. We believe this observation represents a virtual classroom. In total, 51 observations were filtered from the dataset with $n = 1490$.

Variable filtering was also applied. Any variable directly used to calculate our target variable was excluded from further consideration. This step eliminated variables *Number Proficient*, *Percent Proficient*, and *State Average* (*% Proficient*) with 12 input variables remaining.

4.2   Modeling

We employed SAS Enterprise Miner 13.2 to build and analyze our models. Two primary model types were considered, Decision Tree and Logit Regression. Each model type was tested with a variety of splitting rules, variable selection criteria, and variable transformations.

4.3   Decision Trees

A decision tree is a rule-based modeling technique [15]. The basic idea is to split the data set (node in the tree) into subsets (children nodes) based on a splitting criterion on the relationship of the target variable with the input variables. The splitting process continues for each child node until certain condition is met.

Three decision rule models are developed for our study, as listed below.

DT1: Decision Tree with Chi-Square splitting rule, no variable transformation
DT2: Decision Tree with Entropy splitting rule, no variable transformation
DT3: Decision Tree with Chi-Square splitting rule, transformed variables

4.3.1   Chi-Square Splitting

The Chi-square splitting process requires considering a number of split points for each input variable. It is performed on each binary split. The optimal split is selected via the *logworth* value. Bonferroni adjustment is applied to compare *logworth* to $-\log(\alpha/m)$ for a predetermined significance level α and number of comparisons m. The Chi-square splitting process is given in Algorithm 1.

We also applied tree pruning to optimize predictive power and parsimony, and specified max tree depth and min size of tree nodes to improve the run-time performance.

4.3.2   Entropy Splitting

Our second decision tree model (DT2) is based on Entropy splitting rules. Entropy for a given node j is defined as

---

Algorithm 1: Chi-Square Splitting *SplitAttribute(t)*

Input: Tree node t with variables v_i's
Input: α significance level
Input: m number of comparisons
Output: j – the splitting variable
begin
    foreach v_i **do**
        $p_i \leftarrow \chi_i^2(t)$
        $logworth_i \leftarrow -\log(p_i)$
        if $logworth_i > -\log(\alpha/m)$ **then**
            v_i is marked significant
        end
    end
    $logworth_j \leftarrow \max_i(logworth_i)$ for v_i marked
    return j
end

$$H(j) = -\sum_{i=1}^{k} P_i \log_2 P_i$$

where P_i is the posterior probability of the child node i of node j. Essentially, we are interested in pure child nodes or nodes which are as close as we can get to either Above Average = 1 or Below Average = 0. The node being considered for split contains both values and we want the split that maximizes child node purity. When considering our possible splits we calculate the reduction in our impurity index (in this case, Entropy). This calculation takes the form

$$H(j) - \sum_{i}^{r} P_i * H(i)$$

where r is the number of child nodes of j. This process continues until a minimum reduction in impurity is met. We used the default settings in SAS Enterprise Miner for this threshold. Pruning of the candidate tree follows the same procedure as outlined above for the Chi-square selection. Maximum tree depth was set to 6 (7 total levels).

4.3.3 Variable Transformation

Our third decision tree (DT3) includes a variable transformation node prior to the decision tree node. The variables are transformed according to variable type, binary and interval in our data. Binary variables do not require any transformations. Interval variables may, however, fit our modeling process better if they are transformed. The "Best" option in Enterprise Miner was selected that attempts multiple methods. The two primary methods are optimal binning and a power transformation which mimics the so-called Box-Cox power ladder to maximize normality.

4.4 Logit Regression

Logit (or Logistic) regression models [16] are a type of general linear regression with binary target variable. Our target variable is binary. Therefore, a nonlinear regression model is called for. The logit regression model differs from general linear equation in the following way. A general linear regression can be expressed as

$$y = \beta_0 + \sum (\beta_i x_i) + \epsilon$$

where y is a continuous dependent variable and x_i represents independent variables which may be either continuous or binary. In our case the dependent variable is not continuous and the probability of $y = 1$ is estimated:

$$P(y = 1) = \frac{1}{1 + \exp(-(\beta_0 + \sum (\beta_i x_i))}$$

The regression coefficients β_i can be exponentiated to determine the odds of y for a change in x_i. The three logit regression models are listed below.

LM1: Logit model with no variable transformations
LM2: Logit model with variable selection performed by decision tree (Chi-Square)
LM3: Logit model with variables transformed and then selected based on Chi-Square and R-Square

The process of selecting variables for an optimal logit model can be conducted in a number of ways. We employ stepwise selection for all three logit models. The general process is outlined in Algorithm 2.

Algorithm 2: Variable Selection for Logit Reg Models

Input: Data set D with variables v_i's
Input: α significance level (e.g. 0.05)
Output: V – set of variables selected for the model
begin
    $V \leftarrow \{v_j\}$ where v_j has the max(R^2)
    foreach v_i **do**
        $p_i \leftarrow \chi_i^2(D)$
        if $p_i > \alpha$ **then**
            $V \leftarrow V \cup \{v_i\}$
        end
    end
    return V
end

4.4.1 Variable Selection

Our second logit model (LM2) employs a decision tree node before the logit model node to select candidate variables. The decision tree is "grown" via Chi-square splitting in the same manner outlined in Sect. 4.3.1. In this case, however, the tree is allowed to grow until the *logworth* value no longer exceeds the threshold. The tree is not pruned but instead the resulting variables are passed to the logit node for consideration in the model.

Variable selection for model 3 (LM3) is conducted post variable transformation. We employed an R^2 selection method. This step consisted of comparing each variable (including transformed versions) in a forward selection process against a minimum R^2 threshold (0.005).

4.4.2 Variable Transformation

In logit model 3 (LM3), we also include a variable transformation step before variable selection. This step follows the same process outlined in Sect. 4.3.3 which was used to transform variables prior to decision tree model DT3.

4.5 Validation

Due to the small number of observations in our dataset, partitioning the data into training, validation, and test sets would be inappropriate. Instead, we employed k-folds cross-validation with k set to 10. The data set is partitioned into k equal sized random samples. One sample is held as a test set while the others are used as training set to build the models. The process of training and testing is repeated until all k folds have an opportunity to serve as the test set. The results are then averaged and taken as final.

5 Experimental Results

5.1 Final Model Selection

Our model selection process mimicked a bracketed single elimination tournament. Two brackets were formed which consisted of the decision tree models on one side and the logit models on the other. The decision tree models were selected based on the lowest misclassification rate while the logit models were selected based on the lowest average squared error. Once the winning model from each bracket was selected the overall winner was chosen based on lowest average squared error.

Table 5 All decision trees—fit statistics

Model #	Misclassification rate	Ave. squared error
DT2*	0.140268	0.101133
DT3	0.151678	0.108770
DT1	0.159732	0.133661

*Model selected

Table 6 Logit regression—fit statistics

Model #	Misclassification rate
LM3*	0.107184
LM2	0.109252
LM1	0.109252

*Model selected

The best decision tree model was the Entropy model (DT2). Table 5 summarizes the misclassification rates and the average squared error for all three decision tree models on the target variable *above average proficiency*. Misclassification rates are within 2% points of one another. The range of the average squared error was 0.032528.

Among the three logit regression models, the best is LM3, the model with variable transformation and variable selection employed. This outcome is not surprising. We might expect the model with variables optimized via the transformation process to be the best performer. However, the range of the average squared errors for the three models is 0.002068. This is a much smaller range than we found in our decision tree models that is 0.0325. Table 6 summarizes the misclassification rates for each model.

All three logit models perform relatively similar. Transformation had little effect on performance.

The Receiver Operator Curve (ROC) chart of the decision tree models and for the logit regression models are shown in Fig. 3a, b respectively.

The final model selected was the entropy based decision tree (DT2). Based on average squared error, this model performed slightly better than the best logit model (LM3). The average squared error for the decision tree model was 0.101133. The average squared error for the logit model was 0.107184. The difference, 0.006051, is small even when compared to the tight range of average squared error for all three logit models. Nonetheless, the decision tree's performance is better.

The full entropy decision tree model is shown in Fig. 4. It begins by splitting on the percentage of students receiving free lunch. Approximately 46% of the sample is classified at a terminal child node (node 3). This implies that the free lunch variable is a strong predictor of school proficiency.

The next split occurs on percentage of students receiving free or reduced lunch. Again, we see a strong prediction of school proficiency. Of the remaining schools and those with less than 39% of students receiving free or reduced lunch, 96% scored above average MEAP proficiency.

Fig. 3 ROC charts of
decision trees and logit
regression

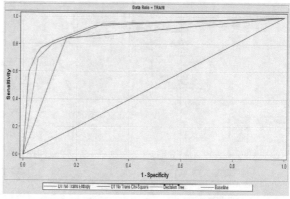

(a) ROC chart – Decision Tree Comparison

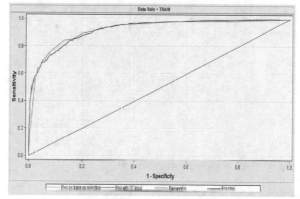

(b) ROC chart – Logit Comparison

The remainder of the 454 schools are split again by free and reduced lunch with
the poorest districts being further defined by jobless rate, share of violent crime
and overall violent crime rate. These results indicate that our model conforms to
known theoretical and empirical research on school performance. The simplicity of
the model also provides ease of explanation to policy makers and stake holders.

5.2 Model Performance

Table 7 reports fit statistics for the winning model (DT2). The ROC index of 0.92 is
well over the common threshold of 0.80 as shown in Fig. 3a. This compares favorably
with the early warning systems reviewed by Bowers et al. The majority of those
models had values below 0.75. In fairness, many of those models were attempting
to predict more complicated outcomes.

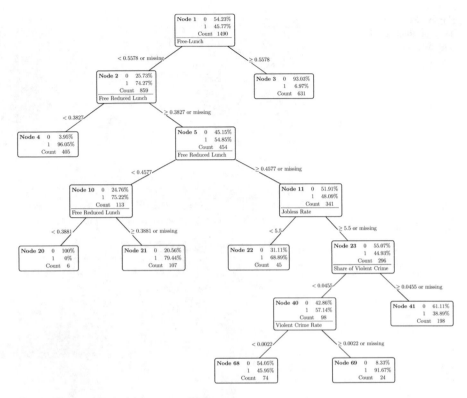

Fig. 4 The entropy decision tree model

Table 7 DT2—fit statistics

Fit statistics	Value
Misclassification rate	0.14
ROC index	0.92
Average squared error	0.10
Cumulative lift	2.10
Gain	109.84

The misclassification rate (0.14) is reasonable but further inspection via a confusion matrix (Table 8) and a misclassification chart (Fig. 5) indicates that our classification of Above Average is less accurate than for schools below the mark. In other words, we have a larger proportion of false positives than we do false negatives. This is a concern. If a school is classified as performing above the average but in reality they are below they may not receive the proper policy prescription.

The goal of the model is predict which schools will fall below the average. Our accuracy in this regard is strong. Less than 10% of schools who are classified as below average are actually above. Again, they are the schools that fall into the false positive category that should be cause for concern.

Table 8 Confusion matrix—DT2

$n = 1490$		Predicted	
		Below average	Above average
Actual	Below average	754	155
	Above average	54	527

Fig. 5 Misclassification chart—DT2

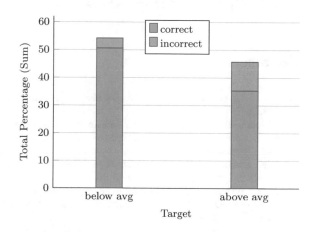

A number of steps may help improve the accuracy of our model. First, the model may be sensitive to the cutoff point. A large number of schools are centered around the mean average proficiency. While this cutoff may be optimal from the state's perspective, it may be not be the optimal point for classification.

We did not employ any interaction terms in any of our models. It may be beneficial to allow, for instance, Big 4 (urban schools) to interact with Charter. Or, perhaps, crime rate interacts with charter schools differently than public schools.

6 Conclusion

Our final model, a decision tree with entropy splitting, does an adequate job of classifying schools as either above or below the state proficiency average. The model compares favorably with other early warning systems in the EDM literature. The tree utilizes a number of socioeconomic variables including percentage of students receiving free lunch, percentage of students receiving free or reduced lunch, jobless rate, violent crime rate, and share of violent crime. The model does not employ any school quality variables such as student-to-teacher ratio, or class size. Class size is traditionally considered a significant factor indicative of student performance, however in recent years research suggests that the characteristics of the students in the classroom is far more important than the number of students in the classroom [17].

Though there is conflicting research regarding this metric, our analysis showed that it is not as important as other Socioeconomic factors.

Additional model types should also be considered in future research. Individual student data is often included in school performance models. The typical approach is a multilevel model approach, one model for the student level and one for the school level. The inclusion of student level data, along with the requisite modeling techniques, may provide further improvement on classification performance.

References

1. Ryan S.J.d. Baker. Data mining for education. *International encyclopedia of education*, 7:112–118, 2010.
2. Alejandro Peña-Ayala. Educational data mining: A survey and a data mining-based analysis of recent works. *Expert systems with applications*, 41(4):1432–1462, 2014.
3. Ryan Shaun Baker, Albert T Corbett, and Kenneth R Koedinger. Detecting student misuse of intelligent tutoring systems. In *International Conference on Intelligent Tutoring Systems*, pages 531–540. Springer, 2004.
4. Joseph E Beck. Engagement tracing: using response times to model student disengagement. In *Proceedings of the 12th International Conference on Artificial Intelligence in Education*, pages 88–95, 2005.
5. R Charles Murray and Kurt VanLehn. Effects of dissuading unnecessary help requests while providing proactive help. In *Proceedings of the 12th International Conference on Artificial Intelligence in Education*, pages 887–889, 2005.
6. Alex J Bowers, Ryan Sprott, and Sherry A Taff. Do we know who will drop out?: A review of the predictors of dropping out of high school: Precision, sensitivity, and specificity. *The High School Journal*, 96(2):77–100, 2013.
7. Bradley Carl, Jed T Richardson, Emily Cheng, HeeJin Kim, and Robert H Meyer. Theory and application of early warning systems for high school and beyond. *Journal of Education for Students Placed at Risk (JESPAR)*, 18(1):29–49, 2013.
8. Brijesh Kumar Baradwaj and Saurabh Pal. Mining educational data to analyze students' performance. *International Journal of Advanced Computer Science and Applications*, 2(6):63–69, 2011.
9. Jared E Knowles. Of needles and haystacks: Building an accurate statewide dropout early warning system in Wisconsin. *Journal of Educational Data Mining*, 7(3):18–67, 2015.
10. Center for Educational Performance and Information. Michigan School Data. http://www.michigan.gov/cepi.
11. Michigan Department of Education. Michigan School Data. https://www.mischooldata.org.
12. Federal Bureau of Investigation. Uniform Crime Report. https://www.fbi.gov/about-us/cjis/ucr/ucr.
13. United States Census Bureau. American Community Survey (ACS). https://www.census.gov/programs-surveys/acs/data.html.
14. Theodore G Chiricos. Rates of crime and unemployment: An analysis of aggregate research evidence. *Social problems*, 34(2):187–212, 1987.
15. J. Ross Quinlan. Induction of decision trees. *Machine learning*, 1(1):81–106, 1986.
16. Frank Harrell. *Regression modeling strategies: with applications to linear models, logistic and ordinal regression, and survival analysis*. Springer, 2015.
17. Caroline M Hoxby. The effects of class size and composition on student achievement: new evidence from natural population variation. Technical report, National Bureau of Economic Research, 1998.

A Development Technique for Mobile Applications Program

Byeondo Kang, Boram Song, Seungwon Yang and Jonseok Lee

Abstract In this paper, we present a development technique, AppSpec, for mobile applications program. The software architecture of business logic for a mobile application can be designed by three models including the application model, the graphic user interface model, and feature model. In order to describe the three models, AppSpec consists of five development phases including requirements analysis, architecture design, navigation design, page design, and implementation and testing. For each phase, we provide diagramming techniques to support the three models for the requirements analysis of the application. Our development procedure helps program developers define functional requirements and design applications architecture through its functional flows. We applied AppSpec to developing a mobile application program, and then presented the products of diagrams as the result of performing development phases of AppSpec.

1 Introduction

Applications programs running on mobile devices are becoming so popular in the business [1]. Mobile applications are different from desktop applications because mobile devices are resource-limited embedded systems and their application

B. Kang (✉) · B. Song
Department of Computer and Information Technology, Daegu University, Gyeongsan, Republic of Korea
e-mail: bdkang@daegu.ac.kr

B. Song
e-mail: br15@naver.com

S. Yang · J. Lee
Department of Computer Engineering, Woosuk University, Wanju County, Republic of Korea
e-mail: yang123@ws.ac.kr

J. Lee
e-mail: jong1007@ws.ac.kr

© Springer International Publishing AG 2017
R. Lee (ed.), *Applied Computing and Information Technology*,
Studies in Computational Intelligence 695,
DOI 10.1007/978-3-319-51472-7_4

47

domain is smaller than desktop domain. A mobile platform consists of various source codes to control a microprocessor and hardware. Mobile applications include various software that use APIs supported by a mobile platform [2]. This paper presents a development technique to help developers understand and define the domain of requirements.

In Sect. 2, we introduce the popular related works for software development process, SPEM(Software and Systems Process Engineering Meta-model) and EPF (Eclipse Process Framework). In Sect. 3, we introduce the characteristics of mobile applications. The software architecture of mobile applications and three models for applications are introduced in Sect. 4. And then we present a development technique, AppSpec, in Sect. 5. We apply our method to an application example in Sect. 6, and the compare our technique with the related works in Sect. 7. Finally we come to a conclusion with brief summary in Sect. 8.

2 Related Works

2.1 SPEM and EPF

The Eclipse Process Framework (EPF) [3, 4] aims at producing a customizable software process engineering framework, with exemplary process content and tools, supporting a broad variety of project types and development styles. The Eclipse Process Framework (EPF) is an open source project that is managed by the Eclipse Foundation. It lies under the top-level Eclipse Technology Project. It has two goals:

- To provide an extensible framework and exemplary tools for software process engineering—method and process authoring, library management, configuring and publishing a process.
- To provide exemplary and extensible process content for a range of software development and management processes supporting iterative, agile, and incremental development, and applicable to a broad set of development platforms and applications. For instance, EPF provides the OpenUP, an agile software development process optimized for small projects.

By using EPF Composer you can create your own software development process by structuring it in one specific way using a predefined schema. This schema is an evolution of the SPEM 1.1 OMG specification referred to as the Unified Method Architecture (UMA). Major parts of UMA went into the adopted revision of SPEM, SPEM 2.0 [5]. The UMA and SPEM schemata support the organization of large amounts of descriptions for development methods and processes. Such method content and processes do not have to be limited to software engineering, but can also cover other design and engineering disciplines, such as mechanical engineering, business transformation, sales cycles, and so on.

2.2 Masam

The MASAM [6] is providing the process based on the Agile Methodology for developing applications software operated on mobile platform. The MASAM gives benefits of rapid development and the reuse of domain knowledge for mobile software. The MASAM methodology is defined as EPF and was distributed to companies for developing mobile applications.

The MASAM has advantage of the SPEM. It can define process according to the context of the mobile company. It supports companies developing mobile applications software. Companies have to tailor MASAM's methodology according to their environments.

3 Characteristics of Mobile Applications

There are two kinds of traditional mobile applications: web applications and native applications [7]. Web-based applications consist of web pages optimized for mobile devices and can be developed by using HTML, JavaScript and CSS. They usually run on a server, so they cannot access the mobile devices' feature (for example, the physical camera device). Native applications are developed for specific mobile devices. They can access the functionalities of mobile devices, such as GPS, file storage, databases, SMS, mail box and etc. They can be downloaded, installed, and sold in applications store. In this paper, mobile applications refer to native mobile applications.

Mobile devices, in particular smart phones, have been popular in our life. Mobile applications are necessary for providing smart phone devices with functionalities for mobile data services. Mobile devices like smart phones are resource-limited embedded systems. And they are developed separately on different development platforms with different operating systems such as RIM of Blackberry, Windows Phone, iOS, Symbian, and Android [8].

Developers for mobile applications have difficulties in handling different devices, multiple operating systems and different programming languages such as Java, Objective-C, and Visual C ++. In addition, mobile applications are developed in small-scale, fast-paced projects to meet competitive market demand [9].

Development of mobile applications program with Android requires Java programming language using the Android SDK which provides the tools and APIs necessary for development [10]. An applications program is packaged as an Android package, "apk" file. An APK file includes all the files related to a single Android application. Figure 1 shows the Android applications architecture.

Fig. 1 Android system
architecture

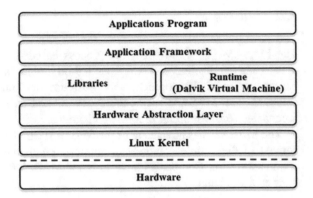

4 Architecture of Applications Program

The mobile application installed on the mobile phone usually consists of two layers: the graphical user interface and the business logic. Figure 2 shows the architecture of mobile applications. The graphic user interface is for user interaction and the business logic is the function of applications program installed the mobile phones. The mobile applications may communicate with a server through a network.

The software architecture of business logic for a mobile application can be designed by three models presented in [11]: the application model, the graphic user interface model, and feature model.

An application model represents the mobile application program. The mobile application program can be represented by the graphic user interface model and the feature model. The graphic user interface model represents screens of mobile phones that users interact to get data services. The screens include the buttons of menu for functions to do the application. The functions are accomplished by clicking the button of menu. The source code for functions is represented by the feature model. The functions for the menu in the screen are represented by the feature model. Figure 3 shows the relationships among three models of an application.

Fig. 2 Architecture of
mobile applications

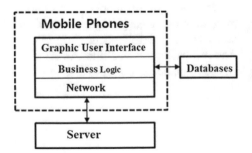

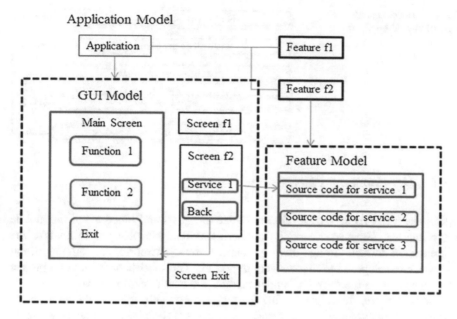

Fig. 3 Three models of an application

In our paper, the application model is described by the use-case diagram, the mind map, and the architecture design diagram presented by AppSpec that are explained in Sect. 5. The graphic user interface model is described by the navigation design diagram and the page detail design diagram. The feature model is source codes for the functions to perform the application.

5 Development Procedure of AppSpec

We propose a development technique, AppSpec, for mobile applications, which includes five phases: requirements analysis, architecture design, navigation design, page design, and implementation and testing. Figure 4 shows the entire development cycle. This procedure is iterative between phases to support feedback. The iterative feedback improves the product quality through recursive review and evaluation. AppSpec is an improved version of the previous method presented in [12, 13].

Fig. 4 Development procedure of AppSpec

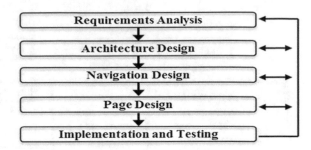

5.1 Development Phases

Requirements Analysis. Developers define the goals and functions of the mobile applications. The purpose of the requirements analysis phase is to analyze the application domain through the viewpoint of users. Therefore, the communication between developers and users is very important. The success or failure of a project is dependent on the degree of understanding the user's requirements.

In this phase, developers define the target users who will use the application. They also analyze the contents and functions, constraints, and who is going to provide the new content. The product of this phase is requirements specification. The specification is described by the Use-case diagram and the Mind-Map.

Architecture Design. Developers determine the most suitable architecture according to the result of the requirements analysis phase. Developers divide the application domains into sub-applications. Well-defined architecture can reduce the complexity of the system and provide the work boundaries for developers. The product of this phase is the architecture design diagram.

Navigation Design. Developers define navigation relationships between pages (screens of smart phones) of the mobile applications. The navigation relationship includes the link relationship and data migration between the pages. The mobile applications program generally consists of more than one page. Users of applications navigate the pages to retrieve some information or to accomplish what they want to do. The product of this phase is the navigation design diagram.

Page Design. Developers design the screen layouts and functions for all of the pages. The pages can be classified into static pages and dynamic pages according to their functions. The function of static pages is to show their contents. The function of dynamic pages is to accomplish tasks such as data processing or accessing databases. The product of this phase is the page detail design diagram. Page detail design diagram consists of the design pattern and the functional flows of each page.

Implementation and Testing. The analysis and design specifications can be implemented in a straightforward manner by programming all of the page detail design diagrams. The behavior of the mobile applications must be tested on the emulator and on the real mobile device because the applications on an emulator may perform differently from running on a real mobile device with various hardware and software versions [14].

5.2 Graphic Notations

Our method provides program developers with two main notations, the component and the connector, for modeling diagrams in each phase of AppSpec. Components represent the functional modules of the system while connectors represent the interactions between components. Figure 5 shows the notations of the diagrams for modeling applications.

Components in the diagrams are classified into the architecture component, the page component, the passive component, the active component, the database component, the group component, and the condition component.

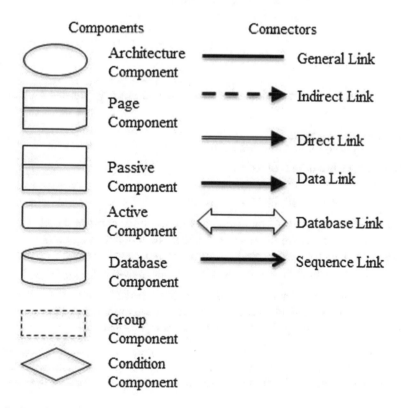

Fig. 5 Graphic notations for design diagrams

The architecture component is used to represent the structure of applications in the architecture diagram. It is a function of an application. The page component represents a page in the navigation design diagram. The passive component represents a static functional module. The active component represents a dynamic module. The database component represents a data repository. The group component can be used to combine a set of components into one group of functions. The condition component is used to specify a condition. All of these components are used in the page detail design diagram.

Connectors in the diagrams are classified into the general link, the indirect link, the direct link, the data link, the DB link, and the sequence link.

The general link represents the existence of any relationships between two components in the architecture design diagram. The indirect link and the data link represent the transitions occurred by a user's clicking on a button. The indirect link does not contain a data transmission between two components. But the data link contains a data transmission between two components. The direct link represents an automatic page link in the program. The Database link represents a data transmission between a functional module and a database. The sequence link represents the sequence of the activation of components. The Database link and the sequence link are used in the page detail design diagram.

5.3   Diagrams

The following five kinds of diagrams are produced after we finish all phases of AppSpec:

- The use-case diagram
- The mind map
- The architecture design diagram
- The navigation design diagram
- The page detail design diagram.

The Use-case Diagram. Application developers must consider the kind of users of the system to be developed. Also they define the functions that users want to do according to their roles. The use-case diagram represents the users of the target system and the functions to be developed within the system to give services to users.

The Mind Map. In order to derive the functions and information for the target system, developers need to extract them from the requirements. The mind map represents the relationships between functions and information.

The Architecture Design Diagram. The architecture of software is defined by computational components and interactions among components. The well-defined structure makes it easy to integrate and maintain the parts of a large application. The architecture design diagram shows the vertical and horizontal structure between functions of applications and does not include the information about the detail algorithms. This diagram is concretized in the navigation design diagram and the page detail design diagram.

The Navigation Design Diagram. The most important characteristic of applications is the navigation feature. Because web applications consist of pages, users of web applications have to explore pages to search for information or accomplish what they want to do.

The navigation design diagram represents the navigation relationships among pages. It shows the link relationships and data transformation between pages.

The Page Detail Design Diagram. The page detail design diagram represents each web page in detail. The pages are classified into static pages or dynamic pages according to their tasks. Some pages may include the characteristics of both.

The static pages display their contents and are described by the design patterns. On the other hand, the dynamic pages perform some tasks and are described by the functional flows to represent the algorithms for the tasks.

6 Application Example: An Application for Ordering a Menu at Inverse-Auction

We applied our method to developing a mobile application that arranges a deal between a user and a company. This application keeps the information about companies by the category of menu. Users can order their menu through the application and then companies can bid their prices for the menu at auction. Users and companies can write reviews and declarations for their deals. A user in this application is a customer who orders a menu. A company is a seller who produces the menu. In case that a user orders a menu, companies that are interested in the order offer their own prices to the user. Then the user chooses one of companies that offer their prices for the menu. Users usually select a company that bids the lowest price for the menu. Figure 6 shows the main screen of the application we want to develop and the development environment for the application.

The first phase of the development procedure in AppSPec is the requirements analysis. In this phase, we develop the use-case diagram and the mind map for the application.

Figure 7 is the use-case diagram that shows the users of the application and their interactions with the application. Users are defined by their roles. In this application, the users are the User and the Company. The roles of the User are to verify their information, to write articles about deal reviews and declarations for the Company, and to make a registration of an auction with the User's location. The roles of the Company are to bid a price of a menu from the User at auction that is registered by

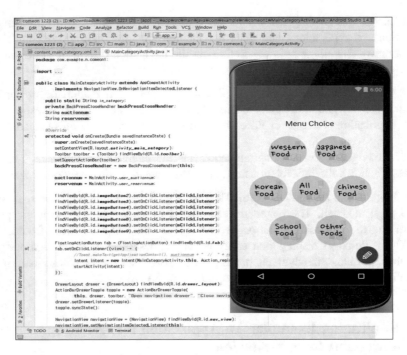

Fig. 6 Application program development environment

Fig. 7 The use-case diagram

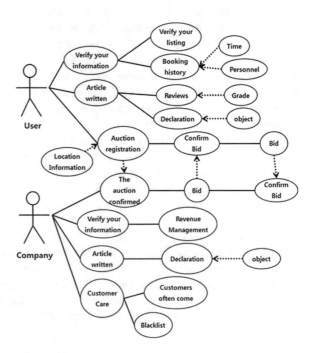

the User, to verify their information, to write articles about declarations about the User, and to keep information about regular customers or black consumers. The User chooses one of companies that offer cheaper prices for the menu.

Figure 8 is the mind map that shows the functions and information of the application. The mind map is extracted and defined from requirements. This application includes the functions for login, user information, business information, category of menu, and auction. The user information consists of user Id, password, name, phone number, and email address. The business information about a company consists of company Id, password, company name, phone number, and email address.

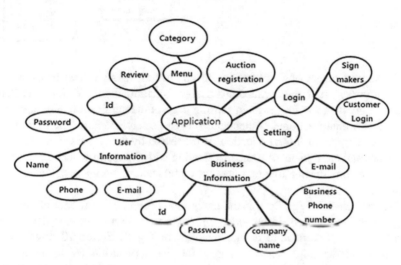

Fig. 8 The mind map

Fig. 9 The architecture design diagram

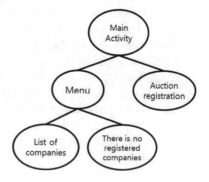

Fig. 10 The navigation
design diagram

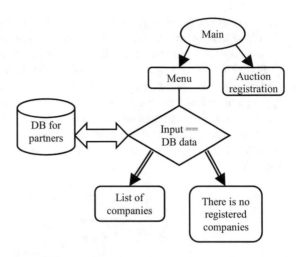

The second phase of AppSpec is the architecture design. The product of the architecture design is the architecture design diagram. Figure 7 shows the architecture design diagram for the main starting page of the application. The application includes two computational components: 'Menu selection' and 'Auction registration'. The component 'Menu selection' is processed by the two components, List of companies and No registered company, dealing with the menu which users ordered. The architecture design diagram includes architecture components and general links.

The third phase of AppSpec is the navigation design. The product of navigation design is the navigation design diagram. Figure 10 shows the navigation design diagram for the architecture design diagram in Fig. 9. Figure 10 represents the navigation relationships between pages for the application by using the page component, the direct link, the data link, and the database link. If any companies dealing with the menu that a user orders exist, the list of companies is displayed. Or the message for no companies is displayed. The component in this diagram is concretized in the page detail design diagram.

After selecting a menu, the component 'Auction registration' is activated for auction. If there are not companies dealing with the menu a user ordered, the message "There are no registered companies" is given to the user. If there are companies dealing with the menu, the list of companies show on screen.

The fourth phase of AppSpec is the page design. The product of this page is the page detail design diagram. Figure 11 shows the page detail design diagram for Fig. 10. Figure 11a is the screen layout of the start page of the application. This screen includes seven image buttons and one icon. Users can choose one of buttons for menu they selected: all food, Korean food, Chinese food, Japanese food, school food, western food, or other food. Figure 11b represents the functional flows of the

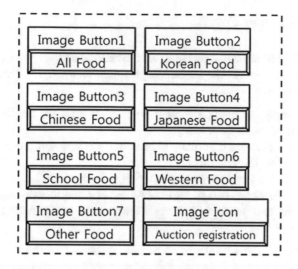

(a) The Design Patterns

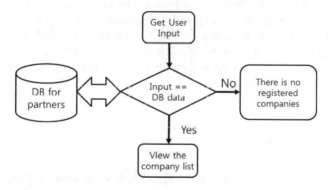

(b) The Functional Flows

Fig. 11 The page detail design diagram, **a** The design patterns **b** The functional flows

function "Menu". It performs the component "View the company list" or the component "No registered companies" after it compares the menu from users with data stored in the DB for companies.

7 Comparison with Related Works

We compare our AppSpec with EDF and MASAM in two aspects of development process and specification method.

7.1   Development Process

EDF and MASAM are based on the SPEM. SPEM is software and systems process engineering meta-model. The methodologies defined through SPEM have the following advantages:

- Clear separation of method content definitions from the development process application of method content
- Consistent maintenance of many alternative development processes
- Many different lifecycle models
- Flexible process variability and extensibility plug-in mechanism
- Reusable process patterns of best practices for rapid process assembly
- Replaceable and reusable process components realizing the principles of encapsulation

In order to get the advantages of SPEM above, all of the development team members must understand the concepts of process assets such as roles, tasks, and work product. It is not easy to define three kinds of process assets. And The processes given from SPEM may be tailored according to project characteristics.

On the other hand, the development procedure of AppSpec consists of five development phases and defines activities in each phases including requirements analysis, architecture design, navigation design, page design, and implementation and testing phase. So it is not general-purpose model, but simple and easy to apply it to mobile applications program.

7.2   Specification Method

EDF and MASAM provide many commercial tools to support their work products. Development teams using SPEM need to learn how to use tools for representing work products. It may be dependent on their process model.

Because AppSpec provides five development phases and work products in each phase including requirements analysis, architecture design, navigation design, page design, and implementation and testing phase, development members only learn the diagramming method in each phase. The diagrams represent the mobile applications by three models such as the application model, the graphic user interface model, and the feature model.

8 Conclusions

There are popular development methodologies for software. But they are general-purpose methodology for the business logics of data services in the real world. So their development procedure needs to be tailored to the mobile applications program.

Mobile applications are different from desktop applications because mobile devices are resource-limited embedded systems and their application domain is smaller than desktop domain. So we present a development technique, AppSpec, that is simply applicable to mobile software running on mobile devices. The development procedure of AppSpec consists of five development phases including requirements analysis, architecture design, navigation design, page design, and implementation and testing phase. It supports three models such as the application model, the graphic user interface model, and the feature model to represent the mobile application programs. The application model is described by the use-case diagram, the mind map, and the architecture design diagram. The graphic user interface model is described by the navigation design diagram and the page detail design diagram. The feature model is source codes for the functions to perform the application.

We applied our method to an application example, and presented documents with diagrams supported by AppSpec. AppSpec helps developers easily understand the requirements of a mobile application, and design its architecture and functional flows through three models. Our work in the future is to implement the graphic editor for supporting the procedures in AppSpec.

Acknowledgements This study was supported by Daegu University grant in 2014.

References

1. H. Muccini, A. Francesco, and P. Esposito, "Software Testing of Mobile Applications: Challenges and Future Research Directions," proceedings of AST, IEEE, 2012, pp. 29–35.
2. Sangwan Cha, Kurz, J. Bernd, Weichang Du, "Toward a unified framework for mobile applications," proceedings of 2009 7th Annual Communication Networks and Services Research Conference, IEEE Computer Society, 2009, pp. 209–216.
3. http://www.eclipse.org/epf.
4. http://en.wikipedia.org/wiki/Eclipse_Process_Framework.
5. http://www.omg.org/spec/SPEM/2.0.
6. Yang-Jae Jeong, Ji-Hyeon Lee, Gyu-Sang Shin, "Development Process of Mobile Application SW Based on Agile Methodology," Proceedings of ICACT, 2008, pp. 362–366.
7. Divya Sambasivan, Nikita John, Shruthi Udayakumar, and Rajat Gupta, "Generic Framework for Mobile Application Development," proceedings of the 2nd Asian Himalayas International Conference on Internet, IEEE, 2011, pp. 1–5.
8. Wei Hu and Hong Guo, "Curriculum Architecture Construction of Mobile Application Development," proceedings of International Symposium on Information Technology in Medicine and Education, IEEE, 2012, pp. 43–47.

9. Mona Erfani and Ali Mesbah, "Reverse Engineering iOS Mobile Applications," proceedings of 19th Working Conference on Reverse Engineering, IEEE Computer Society, 2012, pp. 177–186.
10. M. Butler, "Android: Changing the Mobile Landscape," Pervasive Computing, IEEE, Vol. 10, 23 December, 2010, pp. 4–7.
11. Peter Braun and Ronny Eckhaus, "Experiences on Model-driven Software Development for Mobile Applications," 15th Annual IEEE International Conference and Workshop on the Engineering of Computer Based Systems, 2008, pp. 490–493.
12. Byeongdo Kang, Jongseok Lee, Jonathan Kissinger and Roger Lee. "A Procedure for the Development of Mobile Applications Software." Software Engineering Research, Management and Applications, Studies in Computational Intelligence Vol. 578, 141–150, 2014.
13. Jongseok Lee, Seungwon Yang, Roger Y. Lee and Byeongdo Kang, "Implementation of Delivery Application using a Development Method for Mobile Applications," Proceedings of ACIT2015, ACIS, July 12-16, pp. 137–141, 2014.
14. V. dantas, F. marinho, A. Costa, and M. Andrade, "Testing Requirements for Mobile Applications," proceedings of ISCIS, IEEE, September 14-16, 2009, pp. 555-560.

Interactive Mobile Applications Development Using Adapting Component Model

Haeng-Kon Kim and Roger Y. Lee

Abstract In the reason of the variability which characterizes the context of such environments, it is important that mobile applications are developed so that they can dynamically adapt their extra functional behavior, in order to optimize the experience perceived by their users. In this paper we discuss some of the problems of the current mobile based human management applications and show how the introduction of adaptive CBD (Component Based Development) model provides flexible and extensible solutions to mobile applications. Mobile applications resources become encapsulated as components, with well-defined interfaces through which all interactions occur. Builders of components can inherit the interfaces and their implementations, and methods (operations) can be redefined to better suit the component. New characteristics, such as concurrency control and persistence, can be obtained by inheriting from suitable base classes, without necessarily requiring any changes to users of these resources. We describe the mobile applications frameworks and adaptive component model, which we have developed, based upon these ideas, and show, through a prototype implementation, how we have used the model to address the problems of referential integrity and transparent component (resource) migration. We will show the prototyping applications using our approaches. We also give indications of future work.

Keywords Mobile applications · Frameworks · Adaptive component-based development model · Referential integrity · Mobility · Distributed systems

H.-K. Kim (✉)
School of Information Technology, Catholic University of Daegu, Daegu, Korea
e-mail: hangkon@cu.ac.kr

R.Y. Lee (✉)
Department of Computer Science, Central Michigan University, Mount Pleasant, USA
e-mail: lee@cps.cmich.edu

© Springer International Publishing AG 2017
R. Lee (ed.), *Applied Computing and Information Technology*,
Studies in Computational Intelligence 695,
DOI 10.1007/978-3-319-51472-7_5

1   Introduction

Typically, mobile computing is defined as the use of distributed systems, comprising a mixed set of static and mobile clients [1]. More refined and componentized approaches have also been proposed with new mobile development paradigms [2]. It is widely accepted that building in safety early in the development process is more cost-effective and results in more robust design [3]. Safety requirements result from safety analysis—a range of techniques devoted to identifying hazards associated with a system and techniques to eliminate or mitigate them [3, 4]. Safety analysis is conducted throughout the whole development process from the requirements conception phase to decommissioning. However, currently the approaches for incorporating safety analysis into use case modeling are scarce. The Unifying Modeling Language (UML) [5] is gaining increasing popularity and has become de facto industry standard for modeling various systems many of which are safety-critical. UML promotes use case driven development process [5] meaning that use cases are the primary artifacts for establishing the desired behavior of the system, verifying and validating it. Elicitation and integration of safety requirements play a paramount role in development of safety-critical systems. Hence there is a high demand on methods for addressing safety in use case modeling. Naturally, the development of software applications featuring such a sophisticated behavior is not easy. It has been suggested that while researchers have made tremendous progress in almost every aspect of mobile computing, still not enough has been achieved in dealing with the complexity which characterizes their development [6]. We will show how making the change to a component-based Development system can yield an extensible infrastructure that is capable of supporting existing functionality and allows the seamless integration of more complex resources and services. We aim to use proven technical solutions from the distributed component-based Development community to show how many of the current problems with the Web can be addressed within the proposed model. In the next section, a critique of the current Web is presented, highlighting existing problems in serving standard resources and the current approach for incorporating nonstandard resources. The section entitled the mobile applications frameworks and adaptive component model component design, its aims, component model, and system architecture. The Illustrations section gives an example, describing how particular Web shortcomings can be addressed within the proposed architecture. The remaining sections describe our implementation progress, plans for further work and concluding remarks.

2 Related Works

2.1 Adapting Component Based Mobile Applications Development

Mobile applications modeling expertise requires both domain knowledge and software knowledge. Mobile Applications modeling disciplines are rapidly accumulating in terms of languages, codified expertise, reference models, and automated tools. The areas where such technologies are extensively practiced, the quality features re neither of main concern nor adequately tackled. It is a well-known truth that CBD is important for large and complex systems but why it is important for mobile device applications. It tackles vital concerns such as productivity, high level of abstraction, partitioning of the system development process from the component development process and reusability [5]. Reusability offers a number of advantages to a software development team. An assembly of component assembly leads to a 70% reduction in development cycle time; an 84% reduction in project cost, and a productivity index of 26.2, compared to an industry norm of 16.9. For the development of mass mobile examination system, CBD is a smart method, but due to its explicit requirements such as real time, safety, reliability, minimum memory and CPU consumption, standard component models cannot be used [5]. Rather than, a new CBD methodology is very much needed for the development of mobile mass examination system to deal with its specific requirements.

Mobile applications frameworks development model in this paper is based on component based software development. One of the principles of computer science field to solve a problem is divide and conquer i.e., divide the bigger problem into smaller chunks. This principle fits into component based development. The aim is to build large computer systems from small pieces called a component that has already been built instead of building complete system from scratch. Software companies have used the same concept to develop software in standardized parts/components. Software components are shipped with the libraries available with software (Fig. 1).

While compositional adaptation is restricted to the middleware and application layers, parameter tuning also applies to lower layers, such as the operating system, protocols, and the hardware [11]. For example, at the protocol level, the TCP dynamically adjusts its window to avoid or recover from network congestion. Also, at the hardware layer, adaptations could target ergonomics (e.g., adjust the display brightness), power management (e.g., turn idle network adapters off), etc. An adaptive system can be abstracted by a number of layers. The hardware layer, which includes all hardware devices, is right below the operating system and protocols layer. These layers have the common characteristic that any changes in them affect the whole device. For example, if the display brightness or the processor speed is adjusted, all applications using them are affected. Similarly, changes at the operating system layer also affect the whole device. For example, the Windows Mobile operating system allows the adjustment of the storage versus the program memory

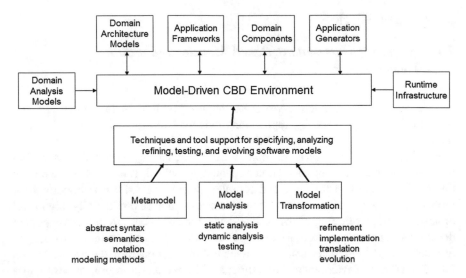

Fig. 1 CBD driven mobile applications development

balance. Clearly, any change in this balance affects the device, and consequently all applications running on it. On the other hand, changes performed at the layers of components and component-based applications have a more limited scope. For example, it is possible to replace a component implementation, or adjust one of its parameters, without causing any direct effect to the applications that are not using it. At the application layer, adaptations are typically achieved with dynamic reconfiguration (classified as changes to the software implementation, composition, or distribution [5]). Finally, besides being limited to these layers, adaptations can also extend beyond the boundaries of a single hosting device. This type of adaptations, which are quoted as distribution adaptations, is of particular interest to this paper. It is argued that users can experience great enhancements in the quality they perceive in their software services if the used devices are capable of synergistically sharing services and components, thus better utilizing the available resources [8].

2.2 CBD Process for Mobile Applications

CBD promises cost-effective productivity assuring a high flexibility and maintenance by assembling the components as independent business processing. The parts. The CBD environment is divided into two Features according to process evolution level. That is, we consider the CBD process as a supply process producing and providing the commercial components into a repository, and consume process supporting component utilization for constructing business solutions [14, 15]. The big picture represents essential works for realizing the CBD process, subjecting the basic principles for component reuse that is acquisition–understanding-applying. CBD process looks different from a traditional one. The development of components, and the composition of an application from the components, are separated. Typically, the two process parts will be executed by different organizations, the component manufacturer and the organization that wants to license and reuse the manufactured components. We refer to these organizations as the component developer as reuse for component and the application composer as reuse with component, respectively as in Fig. 2. Component development is a traditional development process since all the usual lifecycle phases are traversed. The main difference is that the end product is not a complete application. This means that the product is comparatively small, which may make development processes suited to small projects preferable. CBD has rapidly become substantial and interesting field in business applications. Especially, since CBD is primarily used as a way to assist in controlling the complexity and risks of large-scale system development, providing an architecture-centric and reuse-centric approach at the build and deployment phases of development. So now, many vendors and researchers have tried to establish the CBD maturity by involving the following strategies [16, 17]:

(1) Efficient building of individual components,
(2) Efficient building of development solutions of in a new domain effectively,
(3) Efficient adapting a existing solutions to new problems and efficient evolution of sets of solutions.

But, by the lack of standardization and clearness for the CBD approach method, we can't expect a practice benefits in business solutions. So, we need the approach techniques in each step for organizing and practicing the CBD process like a Fig. 3 [6].

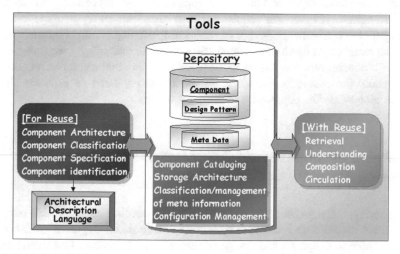

Fig. 2 Basis techniques for mobile applications development using CBD process

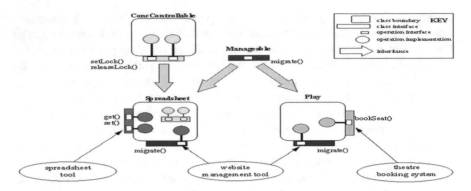

Fig. 3 Illustration of using the adapting component model with interactive mobile applications development component model

2.2.1 Component Interfaces

In an ideal situation, component interfaces would be formally specified, and a CBD would perform formal reasoning to ensure the semantic compatibility of component implementations with their interfaces. However, such reasoning tools are still not widely available or widely used by practitioners, and most commercial components do not have formally specified interfaces. A global namespace of interfaces partly solves the problem of how a CBD will ensure consistency between the semantics of a provided component and the semantics required of the component [18, 19].

While there may be different interfaces providing the same functionality in a global namespace of interfaces, two interfaces with the same name are intended to be functionally equivalent. On a fundamental level, this greatly simplifies the problem of matching provided components to required semantics, since the problem is reduced to name equality. Only when components do not match at the interface level is human intervention required: Either they are truly incompatible (i.e., incompatible on a semantic level), or the incompatibility is only syntactic, so that they can be matched by simple manual adaptation (for example by wrapping one of them). Of course, mechanisms are still needed to ensure that a component correctly implements the semantics promised by its interfaces, but this problem already existed along-side the component matching problem.

3 Design of Interactive Mobile Applications Development Using Adapting Component Model

The primary componentize of our research is to develop an extensible adapting component model with interactive mobile applications development infrastructure which is able to support a wide range of resources and services. Our model makes extensive use of the concepts of component-orientation to achieve the necessary extensibility characteristics. Within this component-Based Development frame-work, proven concepts from the distributed component-Based Development com-munity will be applied to the problems currently facing the AHMS. The interactions between the system components are described in the section entitled "System Architecture," which is followed by a section entitled "Adapting component model with interactive mobile applications development component properties" which classifies and describes a collection of properties applicable to different classes of Adapting component model with interactive mobile applications development Component.

3.1 Interactive Mobile Applications Development Using Adapting Component Model

In common with the current Web, the proposed adapting component model with interactive mobile applications development component architecture consists of three basic entity types, namely, clients, servers, and published components.

In the current Web and mobile environment, these three types correspond to Web browsers (e.g., mosaic), Web daemons (e.g., CERN HTTPD), and docu-mentation resources (e.g., HTML documents) respectively. Our architecture sup-ports both client-component (client-server), and inter component (peer-to-peer) communication.

Figure 3, illustrates the logical view of client-component interactions within the Adapting component model with interactive mobile applications development component architecture. A single server process is shown, managing a single Adapting component model with interactive mobile applications development component (although servers are capable of managing multiple components of different types), which is being accessed via two different clients, a standard Web browser, and a dedicated bespoke application. This diagram highlights *interoper-ability* as one of the key concepts of the architecture, that is, the support for component accessibility via different applications using multiple protocols. As stated earlier, Adapting component model with interactive mobile applications development components are encapsulated, meaning that they are responsible for managing their own properties (e.g., security, persistence, concurrency control etc.) rather than the application accessing the component. For example, in the case of concurrency, the component manages its own access control, based upon its internal policy, irrespective of which application method invocations originate from. The local representation of a component, together with the available operations, may vary depending upon the particular type of client accessing it (Fig. 4).

Although, it has been already stated that we believe CGI to be too low-level for direct programming, CGI interfaces to remote components can be automatically created using stub-generation tools. We have implemented a basic stub-generator, which uses an abstract definition of the remote component, and ANSA have

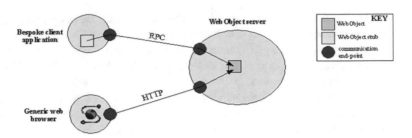

Fig. 4 Client-component interactions for adapting component model with interactive mobile applications development

recently released a more complete tool based on CORBA IDL. Recent developments using interpreted languages within the Web, including Java and SafeTcl are potentially very useful for developing client-side interfaces to Adapting component model with interactive mobile applications development Components. Using such languages, complex, architecture-neutral, front-ends dedicated to a particular Adapting component model with interactive mobile applications development component class can be developed, supporting rich sets of operations.

3.2 Components and Variation Points

A prototype component framework has been implemented, using the Java language [8]. The components are defined as classes, annotated with both required and optional metadata. Furthermore, the framework implements standardized component containers [3], providing runtime support for dynamic adaptations and life-cycle management. Similarly to the majority of industrial component models (e.g., CCM, COM and EJB), the components are considered as similar to object-oriented classes in the sense that they are instantiated and their instances can be stateful. Supporting dynamic compositional adaptations (i.e., through dynamic reconfigurations) is a major research area itself. Kramer and Magee have detected a number of important issues for dynamic reconfigurations, most notably the requirement for quiescence [8]. The component framework defines a set of metadata which are required to enable dynamic adaptations. These metadata are defined inline with the code using annotations. The attached metadata define information such as a unique identifier, a list of roles they implement and a list of roles they export.

The annotation-based mechanisms are used for specifying both the offered and the required roles of components. These roles are then used to facilitate the dynamic composition of applications. The framework achieves dynamic composition by dynamically adding and removing bindings between components, thus enabling the dynamic configuration and reconfiguration of component-based applications.

In this manner, the framework acts as a broker, managing the available and the required roles. Different composition plans are formed by matching required services to offered ones. The actual binding of the components is achieved with the use of reflection, which is a standard feature of Java. With the use of the annotation mechanisms, the components expose both their required and their offered roles. Using these metadata, the components can be connected to each other to form a composition.

Furthermore, hierarchical composition is achieved by implementing the external view of a component through another composition. One of the main advantages of this framework is that it allows the dynamic planning of compositions, as opposed to frameworks which require a predefined set of possible adaptations. While this paper focuses on compositional adaptation, further adaptivity (e.g., at the hardware layer) is also possible (i.e., parameter tuning) [8].

3.3　Adapting Component Model with Interactive Mobile Applications Development

In order to construct adaptive applications, the developers specify how an application should be composed, and when. The first part is achieved with the construction of components with the use of roles and variation points. The latter also requires a mechanism to reason on the context and to select variations. Naturally, these two requirements separate the development phase in two parts: developing the application logic and defining the adaptive behavior. An apparent advantage of this approach is that the same components can be reused for the development of additional, adaptive applications (naturally inherited from the component-oriented approach). Furthermore, the same adaptation strategies can be reused in the context of different applications. For example, a strategy which monitors the network requirements of an application, as a function of its components, can be reused for different applications as well.

Finally, because of the high variability which characterizes mobile environments, it is important that the adaptations can be decided and implemented in a quick and efficient manner (i.e. to cope with frequent and unpredicted disconnections). The following paragraphs describe the two required phases. In addition to client-component communication, our architecture also supports inter-component communication, regardless of the components' location. In effect, the architecture may be viewed as a single distributed service, partitioned over different hosts as illustrated in Fig. 5. Inter-component communication is used for a variety of purposes, including referencing, migration, caching, and replication. In addition to Adapting component model with interactive mobile applications development Components, servers may contain Adapting component model with interactive mobile applications development component stubs, or aliases, which are named components that simply forward operation invocations to another component, transparently to clients. One particular use of aliases is in implementation of name-servers, since a name-server may be viewed simply as a collection of named

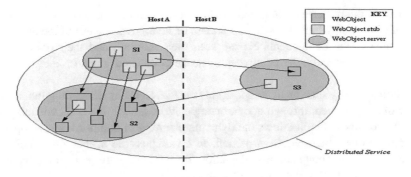

Fig. 5 Inter-component interactions

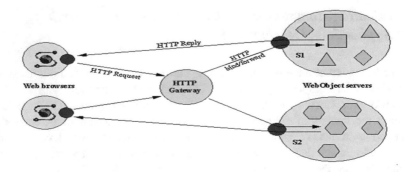

Fig. 6 Client-component communication through gateway

components which alias other components with alternative names (server S1 in diagram). Components may also contain stubs to other components (shown in S2 in diagram). This feature is used in our implementation of referencing, which is described further in the "Illustrations" section.

One method of interfacing with multiple servers is to make use of an HTTP Gateway, which uses stub components to forward component invocations through to the appropriate server. The gateway is transparent to clients accessing the components; incoming requests are simply forwarded to the destination component, which parses the request and replies accordingly. This is illustrated in Fig. 6, in which server S1 manages a number of different types of component (illustrated by different shapes) and server S2 manages components of a single type. As the processing of operations is entirely the responsibility of the individual component, the introduction of new component types is transparent to the gateway.

Based on critiques of the current mobile by ourselves and others [10], and also our experience with distributed systems in general, we have attempted to identify the set of properties that are required by Adapting component model with inter-active mobile applications development Components. We have classified these properties into three categories: core properties, common properties, and class-specific properties. In this section we shall present what we believe to be the core properties required by all Adapting component model with interactive mobile applications development Components and give examples of some common properties.

Four properties have been identified as being the core requirements for adapting component model with interactive mobile applications development components: Naming, Sharing, Mobility, and Referencing. The implementation of these properties is divided between the components themselves and the supporting infrastructure, which manages the components. Each property will be considered in turn.

Naming: One of the fundamental concepts of the component-Based Development paradigm is identity. The ability to name a component is required in order to unambiguously communicate with and about it. Context-relative naming is an essential feature of our environment so as to support interoperability and scalability. As mentioned previously, different clients may use different local representations of a remote component (URLs, client-stub components, etc.). Since it is impractical to impose new naming conventions on existing systems, we require the ability to translate names between system-boundaries. Furthermore, for extensibility, we need to be able to incorporate new naming systems. Within our design, naming is provided via the component infrastructure.

3.4   Implementation of Adapting Component Model with Interactive Mobile Applications Development

Having described our model in the previous sections, we shall now illustrate how two of the core properties, referencing and mobility, are implemented within the model. Our aim is to address the current problem of broken links and provide transparent component migration.

In our model Web resources are represented as Adapting component model with interactive mobile applications development Components and may be referenced from some *root*, either directly, via Adapting component model with interactive mobile applications development component stubs, or by being contained within another Adapting component model with interactive mobile applications development component (note that Adapting component model with interactive mobile applications development component stubs are themselves Adapting component model with interactive mobile applications development Components). This is illustrated in Fig. 7, which shows a number of components, all of which are reachable from some roots. Our service maintains the distributed referencing graph and uses reference counting to detect unreferenced components. Stubs, when created, perform an explicit *bind* operation on the component they refer to (thereby incrementing the component's reference count) and perform an *unbind* operation whenever the stub is deleted (thereby decrementing the count).

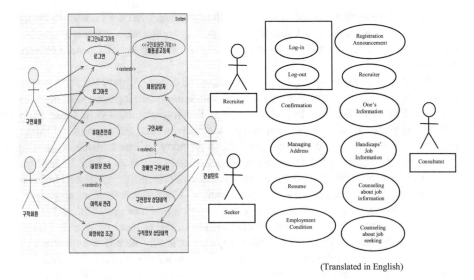

(Translated in English)

Fig. 7 Adapting component model with interactive mobile applications development modeling using UML

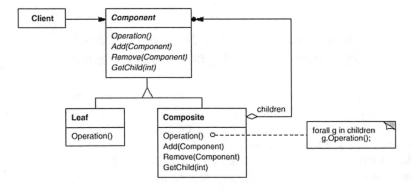

Fig. 8 Possible compositions for the schedule manager application

3.5 Dynamic Composition

At this point, the developer has specified a number of component implementations, along with a set of specifications about the roles they offer and the roles they require. Given these, the framework can plan a set of valid compositions, as illustrated by Fig. 8.

So far, in this phase the developers have defined the set of components, along with their offered and required roles. These artifacts however, cannot result to an adaptive application until the framework is instructed on how and when each composition is selected. As it has been argued already, the task of defining how the application is adapted is a different concern, which should be kept as independent as possible from the task of defining the core application logic. In this respect, it is the responsibility of the second phase to define which composition is more suitable for each context in an independent and reusable manner. The actual evaluation of the utility functions takes place when relevant context changes occur. At that point, the framework evaluates the utility as a function of the new context and properties, and decides whether a new composition can improve on the existing one.

4   Conclusions

The Adapting component model with interactive mobile applications development component model, presented in this paper, is intended to provide a flexible and extensible way of building Web and mobile applications, where Web resources are encapsulated as components with well-defined interfaces. Components inherit desirable characteristics, redefining operations as is appropriate; users interact with these components in a uniform manner. We have identified three categories of component properties: core, common, and specific, and have described an implementation using the core properties which addresses what we believe to be one of the most significant problems facing the current mobile that of referential integrity. A key feature of our design is support for interoperability; for example, in addition to sophisticated clients which may use the rich component inter faces that our model provides, our implementation will also allow Adapting component model with interactive mobile applications development Components to continue to be accessed using existing mobile browsers.

We also describe the mobile applications frameworks and adaptive component model, which we have developed, based upon these ideas, and show, through a prototype implementation, how we have used the model to address the problems of referential integrity and transparent component (resource) migration. We will show the prototyping applications using our approaches. Indications were given on future work.

Acknowledgements This research was Supported by the MSIP (Ministry of Science, ICT and Future Planning), Korea, under the C-ITRC (Convergence Information Technology Research Center) support program (IITP-2016-H8601-16-1007) supervised by the IITP (Institute for Information & communication Technology Promotion).

This research was also supported by the International Research & Development Pro-gram of the National Research Foundation of Korea (NRF) funded by the Ministry of Science, ICT & Future Planning (Grant number: K 2014075112).

References

1. Satyanarayanan, M.: Pervasive Computing: Vision and Challenges, IEEE Personal Communications, vol. 8, no. 4, pp. 44–49, (2001) August.

2. Czyperski, C.: Component Software: Beyond Object-Oriented Programming, ACM Press/Addison-Wesley, (2008).
3. OMG, Common Component Request Broker Architecture and Specification, OMG Document Number 91.12.1
4. Ogbuji, U.: The Past, Present and Future of Mobile Services, http://www.Mobile services. org/index.php/article/articleview/663/4/61/, (2004).
5. Barnawi, A., Qureshi, M. R. J., Khan, A. I.: A Framework for Next Generation Mobile and Wireless Networks Application Development using Hybrid Component Based Development Model", International Journal of Research and Reviews in Next Generation Networks (IJRRNGN), vol. 1, no. 2, pp. 51–58, December (2011).
6. Litoiu, M.: Migrating to Mobile Services-latency and scalability, Proceedings of Fourth International Workshop on Mobile Site Evolution, pp. 13–20, October (2002). URL: http:// www.tigris.org/
7. Brown, A.: Using service-oriented architecture and component-based development to build Mobile service applications," Rational Software white paper from IBM, (2002). 4.
8. Paspallis, N., Papadopoulos, G. A.: An Approach for Developing Adaptive, Mobile Applications with Separation of Concerns, Proceedings of the 30th Annual COMPSAC'06, (2006).
9. Soley, R. and OMG Staff Strategy Group.: Model Driven Architecture, OMG Whit Paper Draft 3.2, at URL: http://www.omg.org/~soley/mda.html, (2000).
10. Poole, J. D.: Model Driven Architecture: Vision, Standards and Emerging Technologies," European Conference on Object-Oriented Programming, at URL: http://www.omg.org/mda/ mda_files/Model-Driven_Architecture.pdf, (2004). 4.
11. Qureshi, M. R. J.: Reuse and Component Based Development," in Proc. of Int. Conf. Software Engineering Research and Practice (SERP'06 Las Vegas, USA), pp. 146–150, 26-29 June (2006).
12. Champion, M., Ferris, C., Newcomer, E., Iona, Orchard, D.: Mobile Services Architecture: W3C Working Draft," http://www.w3.org/TR/ws-arch/, (2002).

Development of Guiding Walking Support Device for Visually Impaired People with the GPS

Tsubasa Sugimoto, Shota Nakashima and Yuhki Kitazono

Abstract Nowadays, when many of the visually impaired people go out, they are using white cane or a guide dog. However, these walking support tools must often rely on the user's senses and free life of the visually impaired people are not yet guaranteed completely. Therefore, in recent years, the development of the walking support device with electronics have been actively carried out. But it is not popular yet in general. One of the reasons is that navigation type walking support device is likely to be heavy equipment, hence the visually impaired people don't use them with ease. For such reason, we conducted development and experiment of the walking support device that enable that they go out with ease by being equipped with sensors to detect obstacles and omni-wheels and navigate them in this study. This device unites the advantages of both electronic white cane and guide dog robot.

1 Introduction

1.1 Present Situation of the Visually Impaired People

Nowadays, serious aging society problem is in progress and people with some failure are increasing. According to the survey by the Ministry of Health, Labor and Welfare in Japan, the number of handicapped people in 2011 is 3.937 million people, it has been increasing year by year. More than half the number of people with disabilities are 65 years of age or more, as you see Fig. 1. 31.6 million people that is 8.2% of

T. Sugimoto · Y. Kitazono (✉)
Department of Creative Engineering, National Institute of Technology,
Kitakyushu College, Kitakyushu, Japan
e-mail: kitazono@kct.ac.jp

T. Sugimoto
e-mail: d31571ts@apps.kct.ac.jp

S. Nakashima
Graduate School of Science and Engineering, Yamaguchi University, Ube, Japan
e-mail: s-naka@yamaguchi-u.ac.jp

© Springer International Publishing AG 2017
R. Lee (ed.), *Applied Computing and Information Technology*,
Studies in Computational Intelligence 695,
DOI 10.1007/978-3-319-51472-7_6

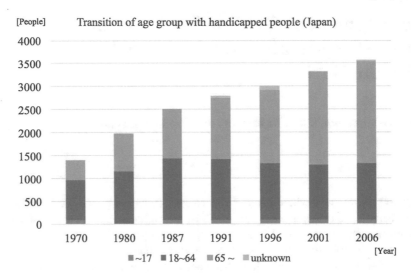

Fig. 1 Transition of age group with handicapped people (Japan)

them are visually impaired people, and it is expected to increase further in the future due to the population aging [1].

1.2 Problems of the Conventional Walking Support Tools

When many of the visually impaired people go out, they have to use a guide dog or a white cane. However, the use of these tools have many problems. First, we will express the problems of when to use the guide dog. Conditions of the visually impaired people using a guide dog are 1. To become proficient at walking training and sense training, 2. That there is no significant gait disorder in good health, 3. To have normal sense of equilibrium and sense of direction, 4. To have ability to hear accurately direction of the sounds, 5. If you still have vision, regardless of the inborn or posteriority, to have only a sense of light or to have only vision ability to recognize the shape of things or that the field of view is too narrow for people to walk, 6. That personality is wholesome, 7. To have a life design to use a guide dog, 8. To like the dog and be able to take care of the dog, 9. That person and the family wants the use of guide dogs. that it is. The visually impaired applying all of these are rare, and the visually impaired who can effectively use a guide dog is one out of six in the United Kingdom. In addition, the training of a guide dog takes more than one year, all dogs trained not become a guide dog. Furthermore, the period during which acts as a guide dog is not only about 8 years. In fact, about 3,000 visually impaired people in Japan need guide dog. In contrast, the number of working dogs are approximately thousand. Second, the problems of when to use the white cane are described. Walking to the braille block or without braille block is difficult because they walk as they are depending on the braille block. Moreover, since they recognize the braille block as they are swinging left and right, the trouble is sometimes caused by the contact of the white cane. some cases develop incidents. Elderly with a visual impairment cannot even possible to use a white cane, therefore they go out with their attendance or many visually impaired people do not all go out. The visually impaired people use their sense of direction to walk and it is impossible to get to a facility which is located at a place with which the user is unacquainted, if they use these two tools. In order to solve these problems, it is urgent need to robotize walking support tool.

1.3 Robotize Walking Support Tool

On the premise that role of the guide dog are 1. To stop in front of the step of stairs, 2. To stop at an intersection, 3. To avoid obstacles, to support the walking, 4. To walk along the edge of the road.

In contrast, functions of the white cane is 1. To ensure the safety of the user by the defense in front of the obstacles and danger, 2. To collect the information necessary to walk such as steps, sidewalk and braille block, 3. Reminder to healthy subjects who are in the same space, It is need that these functions implement and add function solving as the problems mentioned before.

2 Related Work

Since the 1960s, even though the research of a walking assisting device has been actively carried out, because of the problem they haven't become common up to now. The fundamental structure is that the ultrasonic sensor and the light sensor support environment recognition in walking by informing information such as around obstacle detected. Initially research, these devices are that to combine the guide dog or white cane and utilize is principle. Therefore, the guide dog and white cane was called primarily support tools, the walking support device was called secondarily support tools.

However, it has also been increasing things that can walk with just walking support device without the guide dog and the white cane in current study. There are many models such as eyeglass-type, Flashlight-type, cane-type and guide dog robot type.

The walking aid robots of current situation, there are those which guiding by recognizing the braille block by using the image processing, avoiding the obstacle by using a sensor such as an ultrasonic sensor and a laser range sensor and guiding by the use of a GPS and self-location estimation technology. These electronic walking assisting device is classified into two [2–6].

One is what is mechanized conventional guide dog or white cane, it is only to support to walk with the senses of the users mainly.

The other lead the user by using such as GPS and image processing, and support for users to walk actively.

Passive support, such as in the former case, users who can use is limited and their action range is restricted. Therefore, it is necessary to develop active walking support robot such as the latter case. However, because a number of active walking support robots that have been developed up to now is a large size, there is a problem that it is difficult for users to go out with ease. Since the 1960s, even though the research of a walking assisting device has been actively carried out, because of the problem they haven't become common up to now.

Chandrakant 'CK' Isi et al. have developed a smart white cane [7]. This is an extension of the environment cognitive function of cane by attaching the ultrasonic sensor to the cane. However it is very good in environment cognition, it must often rely on the user's senses and trouble caused by the contact of the white cane has not been able to solve.

Furthermore, in 2011, in cooperative with Akita Seiko Co., Ltd. and Akita Prefectural University, the electronic white cane is developed and commercialized. This is what that detects obstacles in the ultrasonic sensor and inform the user by vibration. This product is a 270 g and lightweight, because a conventional white cane is almost the same shape, users can go out with ease. However, this is merely somewhat to solve the problem of wielding white cane, cannot exceed the range of the walking support functions of white cane.

Electronic walking support machine using the omni-wheel and cane have also been developed. The guide cane developed by Johann Bordstein et al. comprises of a long handle and "sensor head" unit that is attached at the distal end of the handle [8]. The sensor head is mounted on a steerable but unpowered 2-wheeled steering axle. This is a positive walking support machine which we aim, up and down the stairs is considered to be difficult because of their weight and shape. Jian Huang et al. Developed a walking support machine using an omni-wheel for the elderly [9]. Because this is not aimed at weight reduction, it is to support the walk to help with a large omni-wheel.

In this study, we propose a walking assisting device which can go out easily while performing active walking assisting and equipped with a large walking support device and the middle of the function of the electronic white cane.

3 Proposed Method

Walking support device developed in this study is structure which implement the wheels to the handle of the cane type. The component drawing is shown in Fig. 2. The form which implement wheels to the cane enable such as this that users can walk passively like they use the guide dog. Furthermore, because the device lead users different from the conventional electric white cane, they can use in a state of feeling secure.

Fig. 2 The structure
diagram of the walking
assisting device

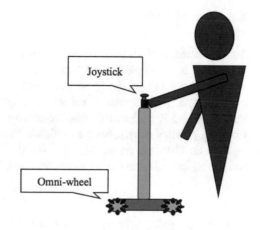

3.1 Drive Unit

3WD48mm omni-wheel robot platform by "Nexus robot" is used as driving part
of this device. This is a platform set of omni-directional movable robot using three
omni-wheels The omni-wheel, it is a unique wheel can be generated a driving force
only in one direction by positioning a plurality of free rollers on the outer circum
ference of the wheel. We use the omni-wheel which are 48 m in diameter. This is
relatively small as for outdoor. Therefore, users don't even give over low step of
sidewalks. They are able to use a method of using this device by lifting it like the
white cane instead, because it is very light. To enable the move to all directions by
placing three omni-wheels every 120°. The three omni-wheel are adopted as a drive
unit, because it is lightweight and possible omnidirectional movement and can control easily. It shows the appearance in Fig. 3.

Fig. 3 3-wheel overview of
omni-wheel robot

3.2   Input Unit

2-axis joystick that visually impaired people can input the direction and speed intuitively has been adopted as the input controller. If the user tilt joystick to direction which they want to go, the device proceed to the direction. Arduino is used as a control for the microcomputer. A schematic diagram of the system is shown in Fig. 4. Further, Fig. 5 is a schematic diagram during navigation. Analysis of the input data from the joystick is performed as follows. We read the angle and magnitude of the input data. Then, we set advancing to 0°, the right 90°, reverse 180°, the left 270°. Relationship angle and the traveling direction is shown in Fig. 6.

3.3   Determination of the Motor Output

Since omni-wheels are set as Fig. 7, outputs of motors are determined to proceed a device in the direction of the vector v. We define output ratio of each motors as $M1$, $M2$, $M3$, and the input as θ, formulas for the output are determined as follows.

$$M1 = -\sin\theta \tag{1}$$

$$M2 = 0.5\sin\theta - \frac{\sqrt{3}}{2}\cos\theta \tag{2}$$

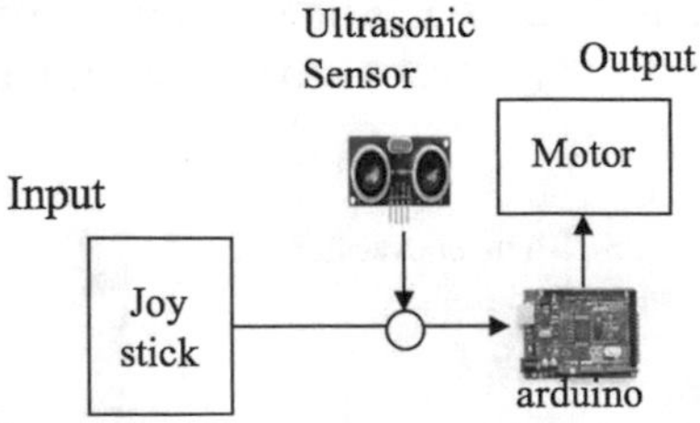

Fig. 4 Schematic view of the structure

Fig. 5 Schematic view of the structure during navigation mode

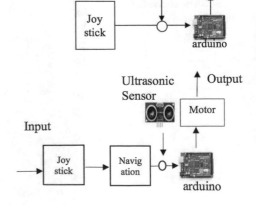

Fig. 6 The angle of the move direction

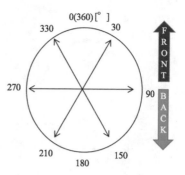

Fig. 7 Vector of the wheels and the move direction

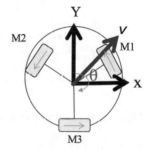

$$M3 = 0.5 \sin \theta + \frac{\sqrt{3}}{2} \cos \theta \qquad (3)$$

Positive values are clockwise, a negative value are anti-clockwise. Thereby, the output torque ratio of each motors when proceeding in the direction of each angle is determined. Figure 8 is graph which is the output value of the motor when move to each direction.

3.4 Obstacle Detection Unit

The ultrasonic sensor is equipped to detect obstacles. Three ultrasonic sensors are fit at center, left and right of motor driving unit. It is shown in Fig. 9. If sensors fit at right and left detect obstacles within 50 mm, the device proceed to each direction avoiding obstacles. In addition, if sensor fit at center do it within 100 mm, it stop. The image illustration is shown in Figs. 10 and 11.

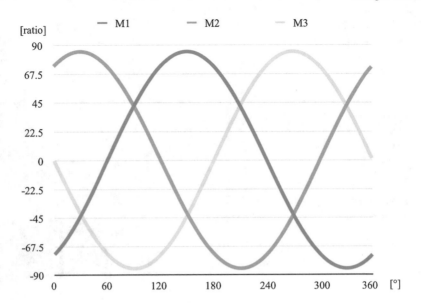

Fig. 8 Output ratio of each motor

Fig. 9 Illustration of
ultrasonic sensor

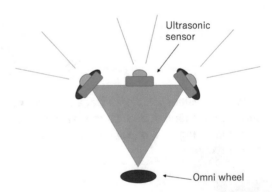

Fig. 10 Left of obstacle

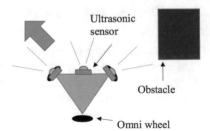

Fig. 11 Right of obstacle

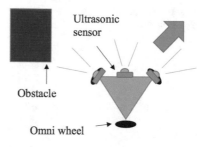

Fig. 12 Detection for down stairs

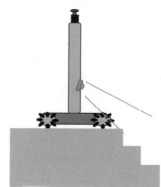

3.5 Stairs Sensor

The ultrasonic sensor for detection stairs is also equipped on cane. It is shown in Fig. 12. This sensor measure the distance to ground. When you are close at down stairs, the distance become long. This is how this device detect the down stairs.

3.6 Navigation Unit

We conduct navigation with GPS module. We conducted precision experiment of GPS module to confirm whether it adequate to navigate. This experiment is conducted by walking the outer periphery of the school while measuring longitude and latitude. Though part of the southwest side is covered in the trees, the environment where measure in this experiment is basically clear place. The result of experiment is shown in Fig. 13. It shows an overlay the results with the actual map to Fig. 14. Very precise measurement values were obtained in clear environment where there aren't high buildings which is obstacles. In case of the environment covered in trees, though some errors occur, it is the range which don't interfere. It is useful that navigation with GPS module from this result of experiment. However navigation can

Fig. 13 Result of
experiment

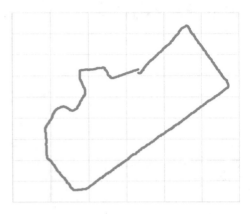

Fig. 14 Result of
experiment overlaid on map

be only used outdoor because GPS date can't be obtained at all in the fully shielded
location such as in a building.

The mechanism of navigation is described. Specification of navigation is to navigate the previously registered route. The user walk having GPS module to regis-
ter, the longitudes and latitudes are recorded to the SD card. As a prerequisite, this
record is performed by the sighted people. Then, the record date is compared with
the present place, the direction is determined. The registered root is reproduced by
repeating it.

3.7   The Length and the Weight

The length of the cane is 1 m, its material which is the vinyl chloride resin is used
for the lightweight. Because it is attached to the three-wheeled robot, overall height
is 1.1 m. The weight is 1294 g. This weight is much lighter than conventional guide
dog robot. This enable to go out with ease, which is one of the most important goal.

Fig. 15 The appearance of
the completed device

Although the conventional guide dog robots cannot climb of stairs, because of this light, this device can go up and down stairs along the railing by lifting. In addition, because it is small compared to the guide dog robots, this can run also in the room of the narrow passage.

3.8 Human Interface

The appearance of the completed device is shown in Fig. 15. Joystick is attached at the tip of the cane, and the mobility is given by that the user hold on it. This device is safety not for the user to be pulled, because this device stop if the user release the hand from joystick. Moreover the user can know unevenness of the road surface to feel the vibration of the machine because wheels run on ground. Thereby the user can stand ready for unevenness of the road surface. Since the device stop physically at step which cannot get over, it is necessarily that the user lift and get over.

4 Experiment

First, we use this device indoor to confirm impression from use of it. Even if it is shielded the visual, this is able to move to direction wanting to go freely because the direction is input with joystick instinctively. Detection of obstacle in right, left and front and avoidance can be conducted smoothly in the point avoidance obstacle. This device don't scare because there are security feeling that if I release the hand, it stop.

Second, we conducted experiment to confirm performance of navigation. The outer periphery of the school was registered like as precision experiment and navigation were conducted in this experiment. The device was used while shielding the visual with eye mask.

Correct navigation was conducted in the clear place but it is conducted to meander in the place where there is an obstacle in the over, such as trees and roof. In addition, machine does not operate as expected in place of slope. Though there are like this problems, we can use sense like as when sighted people lead as the navigation robot.

5   Conclusion

In this study, we proposed a new form of the walking support device which navigate, and developed it. Peculiarity when compare this walking support device with conventional electric white cane or the guide dog robot not to only use with ease like electric white cane but to have the guidance function like the guide dog robot. Thus, we can develop active walking support device using easy and having advantage of both electric white cane and the guide dog robot.

This experiment was conducted in the place where obstacles is comparatively less, and almost no traffic and traffic light. However, There are many place where is traffic, traffic light and obstacle in environment using navigation such as in city. It is necessary to detect traffic light and side walk by such as image data processing to manage it. Furthermore, problem of slope is solved by being equipped with sensor detecting inclination. In the future, we would carry out solution of these problems.

References

1. Cabinet Office, Government of Japan, "Annual Report on Government Measures for Peoples with disabilities (Summary) 2007"
2. Kotani Shinji, Kiyohiro Tomoaki, Mori Hideo, "Development of the Robotic Travel Aid for the Visually Impaired", ITE Transactions on Media Technology and Applications Vol. 51, No. 6, pp. 878–885, 1997
3. Takeshi Kurata, Masakatsu Kourogi, Tomoya Ishikawa, Yoshinari Kameda, Kyota Aoki, Jun Ishikawa, "Navigation System for Visually-Impaired Pedestrians—Preliminary Evaluation of Position Measurement and Obstacle Detection -", IEICE Technical Report, Vol. 110, No. 238, pp. 67–72, 2010.
4. API Co. Ltd.: "Smart Electronics White Cane System" http://www.api-kk.com/denshi-hakujo/
5. Kazuo Kawada, Toru Yamamoto, Yasuhiro Mada, Takuya Tsutsui: "Development of an Intelligent White Cane for Visually Handicapped Persons", The JSME Symposium on Welfare Engineering, Vol. 2002, No. 2–34, pp. 241–244.
6. Yoshikazu Seki: "Transition of Assistive Technologies for Mobility of Blind People", IEICE Technical Report, Vol. 111, No. 58, pp. 87–90, 2011.

7. Chandrakant 'CK' Isi, "Affordable SmartCane Uses SONAR To Guide Visually Impaired," unpublished.   http://www.techtree.com/content/features/6768/affordable-smartcane-uses-sonar-guide-visually-impaired.html
8. Johann Borenstein and Iwan Ulrich, "The GuideCane—A Computerized Travel Aid for the Active Guidance of Blind Pedestrians," Proceedings of the IEEE International Conference on Robotics and Automation, NM, Apr. 21–27, 1997, pp. 1283–1288.
9. Jian Huang and Pei Di, "Motion Control of Omni-Directional Type Cane Robot Based on Human Intention," 2008 IEEE/RSF International Conference on Intelligent Robots and Systems Acropolls Convention Center Nlce, France, Sept. 22–26, 2008, pp. 272–278.

User Evaluation Prediction Models Based on Conjoint Analysis and Neural Networks for Interactive Evolutionary Computation

Ryuya Akase and Yoshihiro Okada

Abstract The authors develop the user evaluation prediction models based on conjoint analysis and neural networks for interactive evolutionary computation (IEC) implemented by interactive genetic algorithm and interactive differential evolution. In addition, the facial expression generation system described in this paper simulates user evaluation based on personalized models and generates images of happy faces and sad faces automatically as an example. IEC that can optimize its targets according to the user's preference and sensibility is attracting attention as an interactive personalization method. However, IEC has the problem of user evaluation fatigue because it requires a lot of user evaluations to search the optimum solution. Therefore, interactive systems employing IEC are used with a user evaluation prediction model so that they can reduce a user's load. The novelties of this study are combination of conjoint analysis and large scale neural networks integrated with user evaluation prediction models. Finally, the authors verify usability of the proposed models by performing user evaluation experiments. As a result, the proposed models indicate better prediction accuracy of user evaluation than a previous research using a simple neural network. Also, the personalized models can simulate user evaluation successfully.

Keywords Interactive evolutionary computation · Genetic algorithm · Differential evolution · Conjoint analysis · Neural network · User evaluation prediction

R. Akase (✉)
Graduate School of Information Science and Electrical Engineering,
Kyushu University, 744 Motooka, Nishi-ku, Fukuoka 819-0395, Japan
e-mail: ryuya.akase@inf.kyushu-u.ac.jp

Y. Okada (✉)
Innovation Center for Educational Resource, Kyushu University,
744 Motooka, Nishi-ku, Fukuoka 819-0395, Japan
e-mail: okada@inf.kyushu-u.ac.jp

© Springer International Publishing AG 2017
R. Lee (ed.), *Applied Computing and Information Technology*,
Studies in Computational Intelligence 695,
DOI 10.1007/978-3-319-51472-7_7

1   Introduction

Interactive evolutionary computation (IEC) is a technique to incorporate user evaluation in the optimization processing. The followings are specific advantages of IEC:

1. Personalize applications based on user preferences.
2. Incorporate knowledge and heuristics of users to the system.
3. Aid creativity of users.
4. Provide user-friendly applications that need not special skills and knowledge.
5. Analyze user preferences by using optimized solution.

Typically, IEC is implemented by interactive genetic algorithm (IGA) or interactive differential evolution (IDE) that overrides the fitness function to user evaluation. It allows personalization of target contents without clarification of the user's preference and sensibility. We have already presented the systems using IEC to create computer graphics contents such as motion and room layout [1–3]. In addition, IEC is applied not only for computer graphics but also for hearing aid adjusting, web design and so on [4–8].

On the other hand, IEC has the problem of user evaluation fatigue because it requires a lot of user evaluations in order to search the optimum solution. Therefore, researchers are coping with this problem as well as expanding a field of application. We also have proposed the approaches such as the user evaluation GUI using self-organizing maps to solve this problem [9].

As for the user evaluation prediction with a simple neural network, Ohsaki et al. had presented the interface that displays individuals in the order of descending predicted user evaluations [10]. They employed the 3-layer neural network configured by 18 input layer units, 10 middle layer units and 20 output layer units, and their system predicts user evaluation dynamically using the actual user evaluation user gave in the previous generation. They verified its usability by means of correlation analysis that analyzes correlations between the actual user evaluation and predicted evaluation. As a result, they reported the neural network could predict user evaluations significantly. Nonetheless, its average of correlation coefficients is approximately 0.3, and it can hardly be considered practical. Basically, our study is based on their approach, and we improve the accuracy of user evaluation prediction by using conjoint analysis and large scale neural networks.

2   Interactive Evolutionary Computation (IEC)

2.1   Interactive Genetic Algorithm (IGA)

Genetic algorithm (GA) is a heuristic search algorithm, and it bases on the Darwinian theory of evolution [11]. It finds the optimum solution by generating individuals that can be the optimum solution. Each individual that is in the population

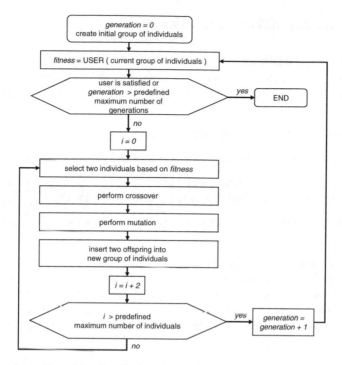

Fig. 1 Flowchart of Interactive Genetic Algorithm (IGA)

develops through the fitness function that determines the ability to solve problems, and crossover and mutation elevate individuals. Interactive genetic algorithm (IGA) replaces the fitness function with the user evaluation to incorporate user preference and knowledge.

Figure 1 shows a flowchart of IGA. The user rates the evolved individuals, and IGA generates new individuals according to the user evaluations. The following are specific procedures:

1. Initialization: This process generates the initial individuals. IGA creates genes configuring an individual randomly.
2. Evaluation: This is the human task. The user rates individuals based on his/her subjective preference. IGA uses these evaluations as the fitness values to evolve the current individuals.
3. Selection: This process selects some highly rated individuals as parents to create a new generation.
4. Crossover: This process transposes gene sequences of two individuals.
5. Mutation: This is a way to change a part of gene in an individual randomly with a fixed probability. It is a useful way to prevent the initial convergence.

Our system iterates these operations (2–5) until the number of iterations exceeds the predefined constant.

2.2 Interactive Differential Evolution (IDE)

Storn et al. proposed Differential Evolution (DE) that is a population-based descent method for numerical optimization [12]. DE has some formats expressed as DE/Base/Num/Cross, and we employed DE/Rand/1/Bin. The following are specific descriptions:

- Base: The selection method of a base vector.
  - Rand: Select a vector from a parent group of individuals randomly.
- Num: The number of difference vectors.
- Cross: The crossover method of a target vector and trial vector.
  - Bin: Use the binomial crossover that performs a crossover with a fixed probability.

The user evaluates individuals in Interactive Differential Evolution (IDE) as with IGA. Figures 2 and 3 illustrate the conceptual diagram and flow chart of IDE. The user selects the target vector or trial vector, and he/she repeats it until the number of evaluated target vectors exceeds the predefined number of individuals to obtain a next generation. IDE reduces the burden of user evaluation because it needs not parallel comparison of all individuals.

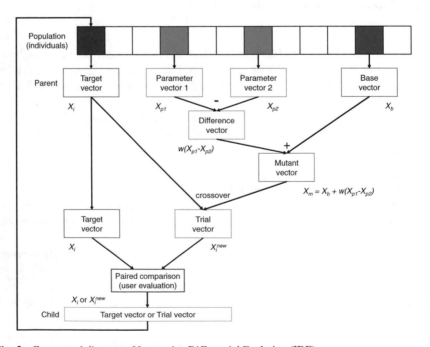

Fig. 2 Conceptual diagram of Interactive Differential Evolution (IDE)

Fig. 3 Flow chart of
Interactive Differential
Evolution (IDE)

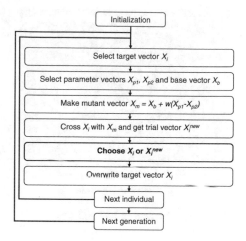

3 System for Experiment

3.1 Overview

Figure 4 illustrates the schematic diagram of our system for experiment. It consists of the server that performs IEC, conjoint analysis and user evaluation prediction using neural networks and web browser to display individuals and to obtain actual user evaluations.

First of all, the system conduct conjoint analysis in order to grasp general user evaluation characteristics (step (a) in Fig. 4). This preprocess provides a simple linear regression model that can estimate user evaluation roughly. We will describe the details of conjoint analysis in Sect. 4.1.

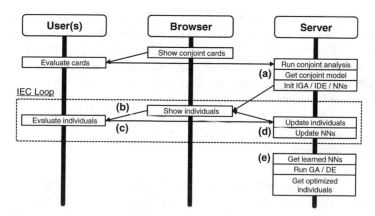

Fig. 4 Schematic diagram of the system for experiment. **a–e** correspond with the symbols in the following figures

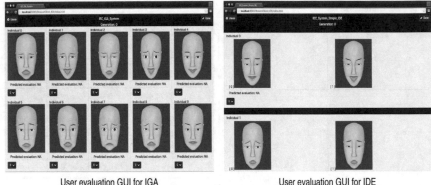

User evaluation GUI for IGA   User evaluation GUI for IDE

Fig. 5 User interface for IGA and IDE

Next, the system executes IEC. As mentioned in Sect. 2, optimized individuals are obtained by repeated proposal of individuals (step (b) in Fig. 4) and user evaluation (step (c) in Fig. 4). In this study, the user creates images of happy faces and sad faces as an example. In addition, the system trains the neural networks in background (step (d) in Fig. 4). We prepare the 3-layer neural network (3NN), 4-layer neural network (4NN), 5-layer neural network (5NN) and pre-trained neural networks that use the training data based on conjoint analysis. The details will be described in Sect. 4.2.

Finally, the system simulates user evaluation based on the trained models (step (e) in Fig. 4) and generates images of happy faces and sad faces automatically. We will describe this result in Sect. 5.

3.2 User Evaluation GUI

Figure 5 shows the web based user interface for IGA and IDE. The browser receives rendered images (individuals) from the server via WebSocket and returns user evaluation values. User evaluation GUI for IGA provides 10 individuals in a generation, and the user grades the individuals within a range of 0–9 according to their sensibility. Similarly, GUI for IDE displays 10 pairs in a generation, and the user selects individuals one by one from the pairs.

3.3 Gene Expression

We use the action units (AUs) implemented by Ekman et al. in order to generate individuals [13]. In particular, our system treats the common AUs listed in Table 1 and Fig. 6 as with Adria et al. [14]. Each AU has the gene index, and each gene element takes the value of 0 (inactive) or 1 (active).

Fig. 6 Action Units (AUs)
to generate individuals

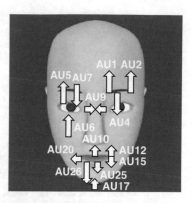

Table 1 Gene index and descriptions of Action Units (AUs)

Gene index	AU	Description
0	1	Inner brow raiser
1	2	Outer brow raiser
2	4	Brow lowerer
3	5	Upper lid raiser
4	6	Cheek raiser
5	7	Lid tightener
6	9	Nose wrinkler
7	10	Upper lip raiser
8	12	Lip corner puller
9	15	Lip corner depressor
10	17	Chin raiser
11	20	Lip stretcher
12	25	Lips part
13	26	Jaw drop

4 User Evaluation Prediction Models

4.1 Conjoint Analysis

Conjoint analysis is an experimental design method that users evaluate some rendering and the system comprehends the importance of each element. Also, conjoint analysis uses conjoint cards that enable integrated evaluation of several factors instead of individual evaluation of separated item. We implemented conjoint cards illustrated in Fig. 7 using 4 features and 2 categories (See Table 2).

The conjoint analysis is calculated by multiple regression analysis. In multiple regression analysis, the relationship between an objective variable (the average score among conjoint cards) and an explanatory variable is represented by the model equa-

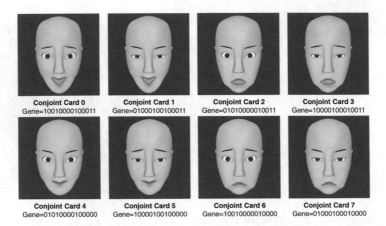

Fig. 7 Conjoint cards the system uses to construct conjoint model

Table 2 Features and categories used in the conjoint analysis. A number means corresponding gene index

Feature	Category I (active)	Category II (inactive)
Eyebrow	1 (AU2)	0 (AU1)
Eyelid	3 (AU5)	5 (AU7)
Lip corner	8 (AU12)	9 (AU15)
Mouth	12 or 13 (AU25 or 26)	12 and 13 (AU25 and 26)

tion. We predict the objective variable using the model equation of multiple regression analysis expressed below:

$$\hat{y} = a + \sum_i b_i x_i \qquad (1)$$

where $\hat{y}$ is an objective variable, x is an explanatory variable, a is a regression constant, and b is a partial regression coefficient. As indicated on Table 2, we used the state of eyebrow, eyelid, lip corner, and mouth as explanatory variables. Note that we need to conduct conjoint analysis before the IEC process to set the linear regression model (conjoint model) that estimates user evaluation.

4.2 Neural Networks

As shown in Fig. 8, user evaluation prediction models using the neural networks use the genes generated by the IEC module as input vectors (x) and adopt actual user evaluations as desired outputs (d). Accordingly, the number of input layer units is

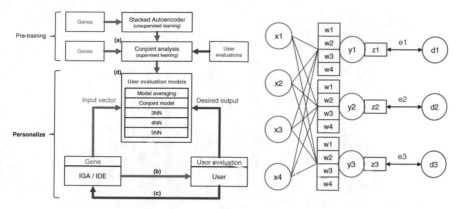

Fig. 8 *Left* Combination of conjoint analysis and large scale neural networks. *Right* Feedforward neural network. x is input (gene), w is weight, $y = \sum_i w_i x_i$, $z = softmax(y)$, e is error function and d is desired output (user evaluation)

the same as number of gene length, and the number of output layer units is equal to the number of individuals in a task of user evaluation.

Our system trains 3-layer, 4-layer and 5-layer feedforward neural network. We defined the number of middle layer units used in those networks empirically. Differentials of the error function are calculated by the back propagation, and the feedforward neural network uses the momentum stochastic gradient descent:

$$w^{t+1} = w^t - \epsilon \nabla E + \mu \Delta w^{t-1} \tag{2}$$

where w is weight, t is time, ϵ is learning rate, ∇E is gradient of the error function, μ is parameter controlling the rate of addition and Δw^{t-1} is a correction amount of the previous weight. In addition, we employed the rectified liner function (unit) as an activation function and the drop out method that avoids the overfitting of feedforward neural network.

Table 3 indicates configurations of the neural networks we adopted. Note that we used the stacked autoencoder in 5-layer neural network to cope with the vanishing gradient problem. The stacked autoencoder takes a neural network apart and performs unsupervised learning applying autoencoder recursively. This method is known to provide good initial weights to a multilayer neural network.

Pre-trained 3NN, 4NN, and 5NN described in the Table 3 perform supervised learning using desired outputs obtained from the conjoint model. In this pre training, our system generates random genes and evaluates the genes based on the conjoint model in order to prepare desired outputs. The pre-trained neural networks are expected to learn general user evaluation.

Finally, we employ model averaging. It is an effective method used in machine learning typically and is known to improve prediction accuracy by means of an average of output values of multiple models. In this study, we use the conjoint model and 3NN model for the model averaging.

Table 3 Configurations of neural networks used in the system for experiment

Model name	Pre training	Input (1st) layer	2nd layer	3rd layer	4th layer	Output (5th) layer	Error function
3NN	–	14 units, DropOut, ReLU	100 units, DropOut, ReLU	–	–	10 units (IGA) or 2 units (IDE), Softmax	Cross entropy
4NN	–	14 units, DropOut, ReLU	100 units, DropOut, ReLU	80 units, DropOut, ReLU	–	10 units (IGA) or 2 units (IDE), Softmax	Cross entropy
5NN	–	14 units, DropOut, ReLU	100 units, DropOut, ReLU	80 units, DropOut, ReLU	60 units, DropOut, ReLU	10 units (IGA) or 2 units (IDE), Softmax	Cross entropy
3NN_pre-trained	Conjoint	14 units, DropOut, ReLU	100 units, DropOut, ReLU	–	–	10 units (IGA) or 2 units (IDE), Softmax	Cross entropy
4NN_pre-trained	Conjoint	14 units, DropOut, ReLU	100 units, DropOut, ReLU	80 units, DropOut, ReLU	–	10 units (IGA) or 2 units (IDE), Softmax	Cross entropy
5NN_pre-trained	Autoencoder, Conjoint	14 units, DropOut, ReLU	100 units, DropOut, ReLU	80 units, DropOut, ReLU	60 units, DropOut, ReLU	10 units (IGA) or 2 units (IDE), Softmax	Cross entropy

5 User Evaluation Experiment

5.1 Experiment 1

In this experiment, 6 subjects create images of happy faces and sad faces interactively using IEC. They also arrange the conjoint cards illustrated in Fig. 7 so that the conjoint card that would match a given concept has a high score. For example, in the case of happy faces, users may arrange the conjoint card numbers such as $1 > 4 > 0 > 5 > 2 > 6 > 3 > 7$.

We use 10 generations and 10 individuals (10 pairs for IDE) the user should evaluates in IEC. In addition, our system simultaneously performs the 3-layer neural network configured by 10 middle layer units (3NN_conventional) Ohsaki et al. had presented and our models described in Sect. 4 in order to compare their prediction accuracy.

The upper side of Fig. 9 shows the average of correlation coefficients between the predictive user evaluation and actual user evaluation in IGA. The average of correlation coefficients of the model averaging, conjoint model and 3NN were 0.4–0.6, and they indicated relevance. On the other hand, the average of correlation coefficients of 3NN_conventional was 0.2, and it could not show adequate relevance.

The lower side of Fig. 9 shows the average of accuracy rate in IDE. Unlike IGA that obtains scores from the user, IDE gets selection results because the user chooses an individual from a displayed pair. Therefore, we observed the accuracy rate indicating the degree of matching between the predictive user evaluation and actual user evaluation. As a result, the model averaging and pre-trained 5NN indicated 0.6 on average, and other models except conjoint model had 0.4–0.5.

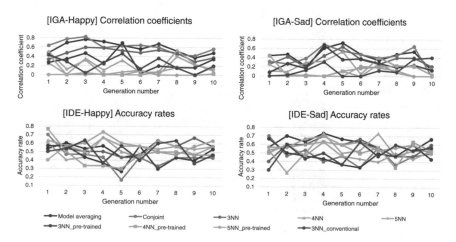

Fig. 9 Result of experiment 1. Conjoint model in IGA, 3NN and their model averaging indicate better prediction accuracy of user evaluation than a previous research (3NN_conventional)

Table 4 Rank based on prediction accuracy of user evaluation

Rank	IGA-Happy	IGA-Sad	IDE-Happy	IDE-Sad
1	Conjoint	Model averaging	5NN pre-trained	Model averaging
2	Model averaging	Conjoint	Model averaging	5NN pre-trained
3	3NN	3NN	4NN	3NN
4	3NN pre-trained	3NN pre-trained	5NN	4NN pre-trained
5	3NN conventional	4NN	3NN	4NN
6	4NN	3NN conventional	3NN conventional	5NN
7	5NN	4NN pre-trained	4NN pre-trained	3NN conventional
8	4NN pre-trained	5NN	3NN pre-trained	Conjoint
9	5NN pre-trained	5NN pre-trained	Conjoint	3NN pre-trained

Table 4 explains the rank based on prediction accuracy of user evaluation. Although there is room for improvement on 4NN and 5NN used in IGA, the model averaging, conjoint model and 3NN indicated better prediction accuracy than 3NN_ conventional. Besides, pre-trained 5NN is considered to be a valid model in IDE.

5.2 Experiment 2

As illustrated in Fig. 10, our system generates images of happy faces and sad faces automatically integrating the personalized models of multiple users. Specifically, it integrates trained 3NNs saved in the experiment 1 using model averaging. The

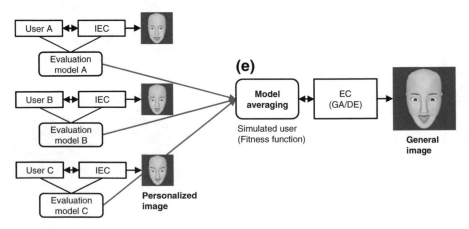

Fig. 10 Integration of personalized models. The system integrates trained 3NNs saved in the experiment 1 using model averaging

GA / Happy   GA / Sad

DE / Happy   DE / Sad

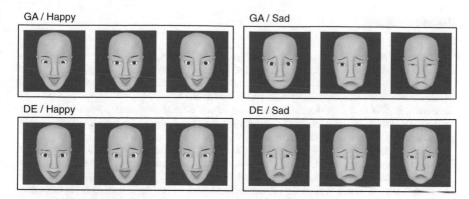

Fig. 11 Result of experiment 2. The integrated model based on personalized models can simulate user evaluation using GA and DE and reproduce the images of happy faces and sad faces successfully

system replaces the fitness function of GA and DE with the integrated model and optimizes individuals by simulating the user evaluation.

Figure 11 demonstrates the results of this experiment. As a result, we were able to confirm that even the 3NN could simulate the user evaluation practically in this experimental condition. This means that we can obtain the images of general happy faces and sad faces automatically if we can save personalized 3NNs during IEC. Model averaging based on personalized models can be also used as a model for pre-training as with the conjoint model. The system can propose adequate individuals to the user by using these models and reduce the fatigue of user evaluation. In addition, there is a possibility that psychiatrists can refer the personalized model in order to inspect personality of an IEC user.

6 Conclusions

The novelties of this study are combination of conjoint analysis and large scale neural networks explained in Fig. 8 and Table 3 and integration of user evaluation prediction models illustrated in Fig. 10. As shown in Fig. 9 and Table 4, the proposed models, especially conjoint model, 3NN and their model averaging indicate better prediction accuracy of user evaluation than a previous research using a simple neural network (3NN_conventional). Also, Fig. 11 describes the personalized models can simulate user evaluation successfully. Therefore, this paper is being expected to contribute to the study that reduces a user's load by simulating user's sensitivity in IEC. As a future work, we plan to improve the prediction accuracy of user evaluation in large scale neural networks so that we can deal with a more complex target.

References

1. Akase, R., Nishino, H., Kagawa, T., Utsumiya, K., and Okada, Y.: An Avatar Motion Generation Method Based on Inverse Kinematics and Interactive Evolutionary Computation, Proc. of the 4th Int. Workshop on Virtual Environment and Network Oriented Applications (VENOA-2012) of CISIS-2012, IEEE CS Press, pp. 741–746, 2012.
2. Akase, R., Okada, Y.: Automatic 3D Furniture Layout Based on Interactive Evolutionary Computation, Proc. of the 5th Int. Workshop on Virtual Environment and Network Oriented Applications of CISIS-2013, IEEE CS Press, pp. 726–731, 2013.
3. Akase, R., Okada, Y.: Web-based Multiuser 3D Room Layout System Using Interactive Evolutionary Computation with Conjoint Analysis, The 7th Int. Symposium on Visual Information Communication and Interaction (VINCI-2014), ACM Press, pp. 178–187, 2014.
4. Takagi, H.: Interactive evolutionary computation: fusion of the capabilities of EC optimization and human evaluation, Proceedings of the IEEE, vol. 89, no. 9, pp. 1275–1296, 2001.
5. Sorn, D. and Sunisa, R.: Web page template design using interactive genetic algorithm, Computer Science and Engineering Conference (ICSEC), 2013 International, IEEE, pp. 206–211, 2013.
6. Mok, T. P., Wang, X. X., Xu, J., and Kwok, Y. L.: Fashion sketch design by interactive genetic algorithms, AIP Conference Proceedings, pp. 357–364, 2012.
7. Ghannem, A., Ghizlane, B., and Marouane K.: Model refactoring using interactive genetic algorithm, Search Based Software Engineering, Springer Berlin Heidelberg, pp. 96–110, 2013.
8. Garcia, H. L., Arauzo, A. A., Salas, M. L., Pierreval, H., and Corchado, E.: Facility layout design using a multiobjective interactive genetic algorithm to support the DM, Expert Systems, pp. 1–14, 2013.
9. Akase, R., Okada, Y.: IGA-based interactive framework using conjoint analysis and SOM for designing room layout, International Journal of Software Engineering and Knowledge Engineering (IJSEKE), Vol. 25, No. 2, pp. 1–35, 2015.
10. Ohsaki, M., and H. Takagi.: Reduction of the fatigue of human interactive EC operators-improvement of present interface by prediction of evaluation order, Journal of the Japanese Society for Artificial Intelligence, vol. 13, no. 5, pp. 712–719, 1998.
11. Dawkins, R.: The blind watchmaker, W.W. Norton, New York, 1986.
12. Storn, R., and Price, K.: Differential evolution - a simple and efficient heuristic for global optimization over continuous spaces, Journal of global optimization, Vol. 11.4, pp. 341–359, 1997.
13. P. Ekman and W. Friesen. Facial Action Coding System: A Technique for the Measurement of Facial Movement. Consulting Psychologists Press, Palo Alto, 1978.
14. Ruiz, Adria, Joost Van de Weijer, and Xavier Binefa. "From Emotions to Action Units with Hidden and Semi-Hidden-Task Learning." Proceedings of the IEEE International Conference on Computer Vision. 2015.

Emotional Video Scene Retrieval Using Multilayer Convolutional Network

Hiroki Nomiya, Shota Sakaue, Mitsuaki Maeda and Teruhisa Hochin

Abstract In order to retrieve impressive scene from a video database, a scene retrieval method based on facial expression recognition (FER) is proposed. The proposed method will be useful to retrieve interesting scenes from lifelog videos. When an impressive event occurs, a certain facial expression will be observed in a person in the video. It is, therefore, important for the impressive scene retrieval to precisely recognize the facial expression of the person. In this paper, we try to construct accurate FER models by introducing a learning framework on the basis of multilayer convolutional network using a number of facial features defined as the positional relations between some facial feature points. The effectiveness of the proposed method is evaluated through an experiment to retrieve emotional scenes from a lifelog video database.

1 Introduction

Owing to recent development of multimedia recording devices such as video cameras and smart phones, people can easily record their daily lives as video data. Since various services to post and view the videos on the Internet are available for free, we can find and/or provide a number of interesting videos at any time. However, some people have a large amount of private video data which cannot be accessed by general public. It will be difficult to retrieve interesting scenes from such private video databases.

In order to solve this issue, we propose a video scene retrieval method to find some impressive scenes from a video database. The proposed method does not require the video database to be public. It thus can be used for private video databases such

H. Nomiya (✉) · S. Sakaue · M. Maeda · T. Hochin
Department of Information Science, Kyoto Institute of Technology, Goshokaido-cho,
Matsugasaki, Sakyo-ku, Kyoto 606-8585, Japan
e-mail: nomiya@kit.ac.jp

T. Hochin
e-mail: hochin@kit.ac.jp

© Springer International Publishing AG 2017
R. Lee (ed.), *Applied Computing and Information Technology*,
Studies in Computational Intelligence 695,
DOI 10.1007/978-3-319-51472-7_8

as lifelog video databases [1]. The proposed method detects the impressive scenes using facial expression recognition (FER). This is because a certain facial expression will be observed in a person in the video when an impressive event occurs. The performance of the video scene retrieval is thus largely dependent on the accuracy of FER.

FER has been applied to video scene detection [2, 3]. Most of existing FER techniques manually define their own facial features and discriminate facial expression using them. The facial feature is one of the core elements in the FER and dominates the recognition performance. It is, however, not easy to manually select good facial features because a variety of very subtle and complex movements of several facial parts will be observed in the appearance of a facial expression.

There is an impressive video scene retrieval method on the basis of an FER method which tries to solve this problem by introducing evolutionary facial feature creation [4]. In this method, useful facial features are generated by combining several arithmetic operations for the positions of facial feature points using genetic programming. This method has an advantage that the facial features are automatically generated. However, it is generally difficult to generate complex facial feature because there are almost infinite combinations of facial features.

The FER technique in the proposed method also utilizes the positional relation between some facial features. Since generating facial features is quite difficult because of the combination problem, the proposed method predefines a number of facial features and selects some useful facial features by introducing a feature selection method. In order to enhance the FER accuracy, we introduce a learning framework on the basis of multilayer convolutional network. Convolutional network is widely used for such as image recognition and speech recognition [5] and currently known as a component of deep learning [6].

The performance of the proposed method is evaluated through an experiment to retrieve emotional scenes from a lifelog video database. We focus on the retrieval of the video scenes with smiles because some interesting events will occur in such scenes and many people will want to retrieve them. The retrieval accuracy of the proposed method is compared with that of the aforementioned retrieval method [4].

The remainder of this paper is organized as follows. Section 2 shows the facial features. Section 3 explains the feature selection method. Section 4 describes the FER method using the facial features. Section 5 introduces the method to detect emotional scenes on the basis of the result of the FER. Section 6 evaluates the performance of the proposed method through an experiment. Section 7 give some consideration about the experimental result. Finally, Sect. 8 concludes this paper.

2   Facial Feature

The proposed method uses a number of facial features computed on the basis of positional relation of several salient points on a face called facial feature points.

2.1 Facial Feature Points

We use 59 facial feature points as shown in Fig. 1. They consist of salient points on left and right eyebrows (10 points), left and right eyes (22 points), a nose (9 points), a mouth (14 points), and left and right nasolabial folds (4 points). They are obtained by using a publicly available software called Luxand FaceSDK (version 4.0) [8].

2.2 Facial Features

The feature value used in the proposed method is computed as the cosine of the angle $(\cos \theta)$ between two line segments formed by three facial feature points p_i, p_j, and p_k as shown in Fig. 2 [9]. Figure 3 shows an example of the facial feature when p_j and p_k are the end points of the mouth and p_i is the center point of the left eye.

Fig. 1 Facial feature points denoted by white squares (this facial image is from Cohn-Kanade AU-Coded Facial Expression Database [7])

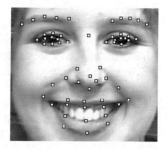

Fig. 2 A facial feature

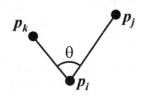

Fig. 3 An example of a facial feature

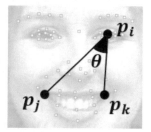

The facial feature value $f_{i,j,k}$ computed from p_i, p_j, and p_k is defined as Eq. (1).

$$f_{i,j,k} = \frac{X_{ji}X_{ki} + Y_{ji}Y_{ki}}{\sqrt{X_{ji}^2 + Y_{ji}^2}\sqrt{X_{ki}^2 + Y_{ki}^2}} \tag{1}$$

Here, X_{ji}, X_{ki}, Y_{ji}, and Y_{ki} are defined as Eq. (2).

$$\begin{aligned} X_{ji} = x_j - x_i, \ X_{ki} = x_k - x_i \\ Y_{ji} = y_j - y_i, \ Y_{ki} = y_k - y_i \end{aligned} \tag{2}$$

where x_a and y_a are the x- and y-coordinates of the facial feature point p_a ($a \in \{i,j,k\}$), respectively.

3 Feature Selection

A total of 97527 possible facial features can be defined from 59 facial feature points.[1] Therefore, using all the possible facial features leads to high computational cost. In addition, there are a lot of useless or redundant facial features. For the purpose of efficient FER, we introduce feature selection to select a small number of useful facial features.

The feature selection is performed on the basis of the usefulness of each facial feature. The usefulness is defined as the variance ratio of the between-class variance to the within-class variance. Equation (3) shows the usefulness Z_i of the ith facial feature.[2]

$$Z_i = \frac{V_i^B}{V_i^W} \tag{3}$$

Here, V_i^B and V_i^W are the ith between-class variance and within-class variance, respectively. They are computed from the facial feature values of training samples. V_i^B and V_i^W are defined by Eqs. (4) and (5), respectively.

$$V_i^B = \sum_{j=1}^{C} \frac{N_j}{N}(\mu_{j,i} - \mu_j)^2 \tag{4}$$

---

[1] One point which is a common end point of two line segments (i.e., p_i shown in Fig. 2) can be selected from 59 facial feature points. Then, two points can be selected from remaining 58 facial feature points. The number of possible facial features is thus $59 \times {}_{58}C_2 = 97527$.

[2] For example, the 1st facial feature corresponds to $f_{1,2,3}$ and the 2nd one is $f_{1,2,4}$. The 97527th facial feature is $f_{57,58,59}$.

$$V_i^W = \frac{1}{N} \sum_{j=1}^{C} \sum_{k=1}^{N_j} (\phi_{i,j,k} - \mu_{j,i})^2 \tag{5}$$

In the above equations, C is the number of facial expressions and corresponds to the number of classes in classification problem. The number of training samples belonging to the ith class is denoted by N_i. The total number of training samples is denoted by N (i.e., $\sum_{i=1}^{C} N_i = N$). $\phi_{i,j,k}$ is the ith facial feature value of the kth sample belonging to the jth class. $\mu_{j,i}$ is the mean value of the ith facial feature values of all the samples belonging to the jth class. μ_i is the mean value of the mean values of the ith facial feature (i.e., $\mu_i = \frac{1}{C} \sum_{j=1}^{C} \mu_{j,i}$).

A larger between-class variance makes the classification problem easier since this indicates that the centroid of each class is distant from each other. On the other hand, a smaller within-class variance is more effective since the samples belonging to the same class are close to each other. Therefore, the facial feature having high usefulness value will be useful to discriminate facial expressions.

4 Facial Expression Recognition

The proposed FER model is constructed based on multilayer convolutional network using the selected facial features. Considering the tradeoff between accuracy and efficiency, we use two layers. The convolutional network for a single layer consists of convolution and pooling.

Similar to the image recognition using convolutional network, we represent the facial feature values of M^2 selected facial features as an $M \times M$ matrix. The M^2 selected facial feature means the top M^2 facial features of the usefulness. The adjacent facial features (i.e., the nth selected facial feature and the $(n + 1)$th one) are sometimes similar to each other (for example, one is calculated from the facial feature points p_1, p_2, and p_3 while the other is calculated from the ones p_1, p_2, and p_4). We represent the selected facial features as a matrix so that diverse facial features are used in the convolution and pooling steps since we believe that the diversity of facial features contributes to higher recognition accuracy.

At the first layer, the convolution operation yields a total of L_1 convolution matrices having the same size as F in accordance with Eq. (6).

$$K_1(i, j, k, l) = \sum_{a=1}^{S} \sum_{b=1}^{S} w_1(a, b, l) F(i, j + a, k + b) \tag{6}$$

$$(1 \leq i \leq N, 1 \leq j \leq M, 1 \leq k \leq M, 1 \leq l \leq L_1)$$

Here, $K_1(i, j, k, l)$ is the (j, k) entry of the lth convolution matrix produced for the ith training sample. $w_1(a, b, l)$ is the (a, b) entry of the lth weight matrix. A weight matrix is an $S \times S$ matrix and S is the patch size. Each entry of the weight matrix is

initialized by a random value generated based on a normal distribution (we experimentally set the mean value to 0 and the standard deviation to 0.1). $F(i, j, k)$ is the
$(j \times M + k)$th (selected) facial feature value of the ith training sample. Note that the
subscripts of K, L and w mean the layer number.

After the convolution operation, a pooling operation is applied to the convolution
matrices. As a result of the pooling operation, a total of L_1 pooling matrices are
produced according to Eq. (7).

$$P_1(i, j, k, l) = \frac{1}{T^2} \sum_{a=1}^{T} \sum_{b=1}^{T} K_1'(i, (j-1)T + a, (k-1)T + b, l) \tag{7}$$

$$\left(1 \le i \le N, 1 \le j \le \frac{M}{T}, 1 \le k \le \frac{M}{T}, 1 \le l \le L_1\right)$$

where, K_1' is defined as Eq. (8).

$$K_1'(i, j, k, l) = RL(K_1(i, j, k, l) + \beta_1(l)) \tag{8}$$

In the above equation, β_1 is an L_1-dimensional bias vector. We initialize the bias
vector such that the values of all the entries are 0.1. The function RL is called ReLU
(rectified linear unit) function and defined as $RL(x) = \max\{0, x\}$. Note that the subscripts of P and β mean the layer number, and that the size of each pooling matrix is
$\frac{M}{T} \times \frac{M}{T}$.

At the second layer, the convolution operation produces a total of L_2 convolution
matrices using the pooling result of the first layer as defined in Eq. (9).

$$K_2(i, j, k, l) = \sum_{a=1}^{S} \sum_{b=1}^{S} \sum_{c=1}^{L_1} w_2(a, b, c, l)P_1(i, j + a, k + b, c) \tag{9}$$

$$\left(1 \le i \le N, 1 \le j \le \frac{M}{T}, 1 \le k \le \frac{M}{T}, 1 \le l \le L_2\right)$$

where, $w_2(a, b, c, l)$ is the (a, b) entry of the weight matrix defined for the cth pooling
matrix generated in the first layer and for the lth convolution matrix generated in the
second layer. We initialize w_2 by the same way as used to initialize w_1. Then, the
pooling operation is performed according to Eq. (10).

$$P_2(i, j, k, l) = \frac{1}{T^2} \sum_{a=1}^{T} \sum_{b=1}^{T} K_2'(i, (j-1)T + a, (k-1)T + b, l) \tag{10}$$

$$\left(1 \le i \le N, 1 \le j \le \frac{M}{T^2}, 1 \le k \le \frac{M}{T^2}, 1 \le l \le L_2\right)$$

where, K_2' is defined as Eq. (11).

$$K'_2(i,j,k,l) = RL(K_2(i,j,k,l) + \beta_2(l)) \tag{11}$$

Note that weight vector β_2 is L_2-dimensional. We initialize β_2 using the same way as in the initialization of β_1.

The output (i.e., a set of pooling matrices) of the second layer is passed to a fully-connected layer. The layer consists of U units and their output values are represented as an $N \times U$ matrix Λ defined as Eq. (12).

$$\Lambda = RL(\bar{P}_2 w_\Lambda + \beta_\Lambda) \tag{12}$$

Here, $\bar{P}_2$ is a matrix defined by converting P_2 into a matrix whose number of rows is N and that of columns is $(\frac{M}{T^2} \times \frac{M}{T^2} \times L_2)$.[3] w_Λ is a weight matrix whose number of rows is $(\frac{M}{T^2} \times \frac{M}{T^2} \times L_2)$ and that of columns is U. β_Λ is an $N \times U$ bias matrix that each row is the same U-dimensional bias vector. When x is a matrix, $RL(x) = X$ such that $X_{ij} = \max\{0, x_{ij}\}$, where X_{ij} and x_{ij} are the (i,j) entries of X and x, respectively. Note that we initialize w_Λ and the bias vector for β_Λ using the same way as used to initialize w_1 and β_1, respectively.

Using the outputs of all units often leads to overfitting. We therefore introduce "dropout" to prevent overfitting [10]. By introducing dropout, several units are dropped at random during training phase. We experimentally set the rate of dropout to 0.5. This means that the outputs of a half of units are ignored when Λ is computed.

Finally, the readout layer is constructed based on the outputs of the fully-connected layer. The output Q of the readout layer is defined as Eq. (13).

$$Q = SM(\Lambda w_Q + \beta_Q) \tag{13}$$

In the above equation, SM is the softmax function defined by Eq. (14).

$$SM(\chi) = \begin{pmatrix} \frac{e^{\chi_{11}}}{\sum_{i=1}^{C} e^{\chi_{1i}}} & \cdots & \frac{e^{\chi_{1C}}}{\sum_{i=1}^{C} e^{\chi_{1i}}} \\ \vdots & \ddots & \vdots \\ \frac{e^{\chi_{N1}}}{\sum_{i=1}^{C} e^{\chi_{Ni}}} & \cdots & \frac{e^{\chi_{NC}}}{\sum_{i=1}^{C} e^{\chi_{Ni}}} \end{pmatrix} \tag{14}$$

where, χ is an $N \times C$ matrix and χ_{ij} is the (i,j) entry of χ. w_Q is a $U \times C$ weight matrix. β_Q is an $N \times C$ bias matrix that each row is the same C-dimensional bias vector. Note that we initialize w_Q and the bias vector for β_Q using the same way as used to initialize w_1 and β_1, respectively.

The output of the readout layer Q is represented as an $N \times C$ matrix. It indicates the possibility of each facial expression for each training sample. For example, the (i,j) entry of Q corresponds to the possibility that the person in the ith training sample expresses the jth facial expression. Therefore, the proposed FER model can predict the facial expression (i.e., class label) of the training examples by Eq. (15).

---

[3]The $(i,1)$ entry of $\bar{P}_2$ is $P_2(i,1,1,1)$, and the $(i,2)$ entry of it is $P_2(i,1,1,2)$, and so on. The $(i, \frac{M}{T^2} \times \frac{M}{T^2} \times L_2)$ entry of $\bar{P}_2$ is $P_2(i, \frac{M}{T^2}, \frac{M}{T^2}, L_2)$.

$$q(i) = \operatorname*{argmax}_{j} Q(i,j) \tag{15}$$

where, $q(i)$ is the predicted class label for the ith training sample and $Q(i,j)$ is the (i,j) entry of Q.

The goal for the construction of the FER model is to optimize the parameters such as weights and biases. To do this, we use Adam algorithm [11]. It is the algorithm for first-order gradient-based optimization of stochastic objective functions. We use the cross entropy as the objective function.

After the training (i.e., the optimization of the parameters), the recognition of the facial expression of unseen samples (test samples) can be performed. This is done simply by replacing the facial feature values of training samples in F (see Eq. (6)) with those of test samples.

5   Emotional Scene Detection

The emotional scenes are detected from a video according to the predicted class label for each frame image. The class label is predicted by the FER model described in Sect. 4. The emotional scenes with a certain facial expression are determined by using the frame images having the corresponding class labels. We make use of the emotional scene detection method proposed in [4].

At the first step of the emotional scene detection, each frame image having the corresponding class label is regarded as a single emotional scene. Then, neighboring emotional scenes are integrated into a single emotional scene. The integration process is repeated until no more emotional scenes can be integrated. The resulting scenes are output as the emotional scenes of the facial expression.

The algorithm of the emotional scene detection is shown in Algorithm 1. Since the emotional scene detection algorithm can find the emotional scenes for a single facial expression, it is required to perform the emotional scene detection C times when there are C kinds of facial expressions in a video.

6   Experiment

6.1   Experimental Settings

6.1.1   Data Set

As the data set to evaluate the proposed method, we prepared six lifelog video clips by six subjects termed A, B, C, D, E, and F. All the subjects are male university students.

Algorithm 1 Emotional scene detection.

Notations:

- E_i^c: The i-th emotional scene in which the facial expression c appears.
- $first(E_i^c)$: Frame number of the beginning frame in E_i^c.
- $last(E_i^c)$: Frame number of the ending frame in E_i^c.
- $length(E_i^c)$: Length of E_i^c. It is equivalent to $last(E_i^c) - first(E_i^c) - 1$.
- $\#int(E_i^c)$: Number of emotional scenes integrated into E_i^c.
- $\#nonemo(E_i^c)$: Number of nonemotional frames in E_i^c. Note that a nonemotional frame means that the facial expression appears in that frame is different from c.
- $dist(E_i^c, E_j^c)$: The distance between E_i^c and E_j^c $(i < j)$. It is equivalent to $first(E_j^c) - last(E_i^c) - 1$.

Initialize:

For each frame image having the class label c, perform the following initialization according to Equation (16):

$$first(E_i^c) = last(E_i^c) = c_i, \ \#int(E_i^c) = 0,$$

$$\#nonemo(E_i^c) = 0, \ length(E_i^c) = 1, \ (1 \leq i \leq M_c) \tag{16}$$

where, c_i is the frame number of the i-th emotional frame in the video. M_c is the number of emotional frames. An emotional frame is the frame having the class label c. That is, each emotional scene consists of a single emotional frame.

Procedure:

1: Find i^* in accordance with Equation (17):

$$i^* = \underset{i}{\operatorname{argmin}} \ dist(E_i^c, E_{i+1}^c)$$

$$s.t. \ dist(E_i^c, E_{i+1}^c) \leq \frac{length(E_i^c) - \#nonemo(E_i^c)}{\#int(E_i^c) + 1}$$

$$\wedge \ dist(E_i^c, E_{i+1}^c) \leq \frac{length(E_{i+1}^c) - \#nonemo(E_{i+1}^c)}{\#int(E_{i+1}^c) + 1} \tag{17}$$

2: If there is no i^* that satisfies Equation (17), finish the procedure and output current emotional scenes. Otherwise, proceed to step 3.

3: Integrate $E_{i^*+1}^c$ into $E_{i^*}^c$ by updating $E_{i^*}^c$ as follows:

$$last(E_{i^*}^c) \leftarrow last(E_{i^*+1}^c), \#int(E_{i^*}^c) \leftarrow \#int(E_{i^*}^c) + 1,$$

$$\#nonemo(E_{i^*}^c) \leftarrow \#nonemo(E_{i^*}^c) + first(E_{i^*+1}^c) -last(E_{i^*}^c) - 1$$

Note that $length(E_{i^*}^c)$ is also updated due to the update of $last(E_{i^*}^c)$.

4: Delete $E_{i^*+1}^c$ and renumber the subscripts of E_i^c so that the emotional scenes become $E_i^c, \ldots, E_{M_c-1}^c$.

5: $M_c \leftarrow M_c - 1$ and return to step 1.

Table 1 Number of samples

Data set	#Samples
A	1585
B	2004
C	1616
D	1361
E	1730
F	1436

These video clips contain the scenes of playing cards recorded by web cameras. A single web camera recorded a single subject's frontal face. This experimental setting is due to the limitation of FaceSDK that it can detect the facial feature points of a single frontal face. While card games are suitable for stably recording frontal faces, a player of most of card games tries to keep a poker face. We thus chose the card games such as Hearts in which the players could clearly express the emotion.

The size of each video is 640×480 pixels and the frame rate is 30 frames per second. Considering the high frame rate, we selected frames from each video after every 10 frames in order to reduce the computational cost. The number of samples (i.e., frames) in each video clip is shown in Table 1. Note that the videos recorded are not shown in this paper because of privacy reasons.

The facial expressions observed in most of the emotional scenes in the video clips were smiles. Thus, we set the value of C (described in Sect. 3) to 2 intending to detect the emotional scenes with smiles, that is, to discriminate smiles and other facial expressions. The ratio of the emotional frames to all the frames varies from 16.6% to 29.6%. A subject is smiling in 24.6% of the frames in the video clip on average.

A two-fold cross validation was used in this experiment by dividing each video clip into the first and second halves. The one was used for the training and the other was used for the test.

6.1.2 Parameter Settings

In the process of the construction of the FER model, several parameters are required for training multilayer convolutional network. We experimentally determined the values of these parameters. The value of each parameter is shown in Table 2.

As described in Sect. 4, we used Adam algorithm for the parameter optimization. This is an iterative algorithm and the number of iterations should be determined. We set the value of iterations to 700 considering the result of a preliminary experiment.

In Adam algorithm, there are four parameters to be experimentally determined [11]. The exponential decay rates β_1 and β_2 are set to 0.9 and 0.999, respectively. Note that they have no relation to β_1 and β_2 described in Sect. 4. The learning rate α and the constant for stability ϵ are set to 10^{-4} and 10^{-8}, respectively.

Table 2 Parameters for FER

Parameter	Value
M	12
S	5
L_1	32
T	2
L_2	64
U	1024

6.2 Experimental Result

The recall, precision, and F-measure of the emotional scene detection are computed for the evaluation of the accuracy of the proposed method. The recall, precision, and F-measure are defined by Eqs. (18), (19), and (20), respectively.

$$recall = \frac{|T \cap \hat{T}|}{|T|} \tag{18}$$

$$precision = \frac{|T \cap \hat{T}|}{|\hat{T}|} \tag{19}$$

$$F-measure = \frac{2 \cdot recall \cdot precision}{recall + precision} \tag{20}$$

where, T is the correct set of emotional frames. One of the authors determined whether each frame was emotional or not prior to the experiment. $\hat{T}$ is the set of emotional frames detected by the proposed method.

We compared the scene detection accuracy of the proposed method with that of an existing method based on an evolutionary facial feature creation [4]. The number of facial features to be selected was set to six for the existing method because it was reported in [4] that six facial features were sufficient for this method. The recall, precision, and F-measure of the proposed method and the existing method are shown in Fig. 4.

7 Consideration

The proposed method outperformed the existing method for four subjects out of six ones in F-measure. In particular, Subject D's F-measure is improved well due to the large improvement in recall. The facial expression of Subject D is relatively weak compared with the other subjects. This makes the detection of his smile more difficult and leads to lower recall in the emotional scene retrieval. There are some people

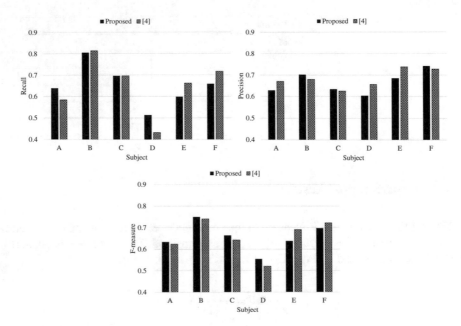

Fig. 4 Emotional scene detection accuracy (recall, precision, and F-measure)

who rarely express strong facial expressions. This result indicates that the proposed method could be suitable for them.

On the other hand, the F-measure of the proposed method for Subjects E and F is lower than that of the existing method. In the proposed method, there are many parameters to be experimentally determined as shown in Table 2, while the number of parameters in the existing method is relatively small. The accuracy for these subjects may be ameliorated by improving the parameter settings.

8 Conclusion

An emotional video scene retrieval method was proposed in this paper. The detection of emotional scenes is performed on the basis of FER. In order to accurately discriminate the facial expressions, we proposed an FER model constructed using several useful facial features and multilayer convolutional network.

The effectiveness of the proposed method was evaluated through an emotional scene detection experiment using some lifelog video clips. The detection accuracy of the proposed method was compared with an existing method. The detection accuracy (F-measure) of the proposed method for two-thirds of the subjects was higher than that of the existing method. The experimental result showed that the proposed method could be effective for the case that accurately detecting emotional scenes was relatively difficult due to weak facial expressions.

Since there are many parameters to be optimized in the proposed method, developing an effective parameter tuning method is required for the improvement of accuracy. This is included in the future work. In the experiment, we focus on only smiles. Evaluating the proposed method using a data set containing a wide variety of facial expressions is also the future work.

Acknowledgements This research is supported by Japan Society for the Promotion of Science, Grant-in-Aid for Young Scientists (B), 15K15993.

References

1. T. Datchakorn, T. Yamasaki, and K. Aizawa, "Practical Experience Recording and Indexing of Life Log Video," Proc. of the 2nd ACM Workshop on Continuous Archival and Retrieval of Personal Experiences, pp. 61–66, 2005.
2. D. Datcu and L. Rothkrantz, "Facial Expression Recognition in Still Pictures and Videos Using Active Appearance Models: A Comparison Approach," Proc. of the 2007 International Conference on Computer Systems and Technologies, pp. 1–6, 2007.
3. G. Fanelli, A. Yao, P.-L. Noel, J. Gall, and L. V. Gool, "Hough Forest-based Facial Expression Recognition from Video Sequences," Proc. of the 11th European Conference on Trends and Topics in Computer Vision, pp. 195–206, 2010.
4. H. Nomiya and T. Hochin, "Emotional Scene Retrieval from Lifelog Videos Using Evolutionary Feature Creation," Studies in Computational Intelligence, Vol. 612, pp. 61–75, 2015.
5. Y. LeCun and Y. Bengio, "Convolutional Networks for Images, Speech, and Time Series," The Handbook of Brain Theory and Neural Networks, MIT Press, pp. 255–258, 1998.
6. Y. LeCun, Y. Bengio, and G. Hinton, "Deep Learning," Nature Vol. 521, pp. 436–444, 2015.
7. T. Kanade, J. F. Cohn, and Y. Tian, "Comprehensive Database for Facial Expression Analysis," Proc. of the 4th IEEE International Conference on Automatic Face and Gesture Recognition, pp. 46–53, 2000.
8. Luxand Inc., Luxand FaceSDK 4.0, http://www.luxand.com/facesdk [September 11, 2016] (current version is 6.1).
9. H. Nomiya, S. Sakaue, and T. Hochin, "Recognition and Intensity Estimation of Facial Expression Using Ensemble Classifiers," Proc. of 15th International Conference on Computer and Information Science, pp. 825–830, 2016.
10. N. Srivastava, G. Hinton, A. Krizhevsky, I. Sutskever, and R. Salakhutdinov, "Dropout: A Simple Way to Prevent Neural Networks from Overfitting," Journal of Machine Learning Research, Vol. 15, No. 1, pp. 1929–1958, 2014.
11. D. Kingma, J. Ba, "Adam: A Method for Stochastic Optimization," Proc. of the 3rd International Conference for Learning Representations, arXiv:1412.6980, 2015.

Proactive Approach for the Prevention of DDoS Attacks in Cloud Computing Environments

Badr Alshehry and William Allen

Abstract Contemporary security systems attempt to provide protection against distributed denial-of-service (DDoS) attacks; however, they mostly use a variety of computing and hardware resources for load distribution and request delays. As a result, ordinary users and website visitors experience timeouts, captchas, and low-speed connections. In this paper, we propose a highly inventive multilayer system for protection against DDoS in the cloud that utilizes Threat Intelligence techniques and a proactive approach to detect traffic behavior anomalies. The first layer of the model analyzes the source IP address in the header of incoming traffic packets and the second layer analyzes the speed of requests and calculates the threshold of the attack speed. If an attack remains undetected, the incoming traffic packets are analyzed against the behavior patterns in the third layer. The fourth layer reduces the traffic load by dispatching the traffic to the proxy, if required, and the fifth layer establishes the need for port hopping between the proxy and the target website if the attack targets a specific web-application. A series of experiments were performed and the results demonstrate that this multilayer approach can detect and mitigate DDoS attacks from a variety of known and unknown sources.

Keywords Distributed denial-of-service attacks · Cloud computing · Proxy firewall · Threat intelligence · Computer security

1 Introduction

Distributed denial-of-service (DDoS) attacks have become highly complicated and have enormous destructive potential. During Q2 of 2015 the most powerful attack occurred at a speed of 250 Gb/s, followed by an attack at 149 Gb/s during Q3 of the

B. Alshehry (✉) · W. Allen
School of Computing, Florida Institute of Technology, Melbourne, FL 32901, USA
e-mail: balshehry2005@myfit.edu

W. Allen
e-mail: wallen@fit.edu

© Springer International Publishing AG 2017
R. Lee (ed.), *Applied Computing and Information Technology*,
Studies in Computational Intelligence 695,
DOI 10.1007/978-3-319-51472-7_9

same year. The total number of DDoS attacks increased by 180% compared to the previous year [1]. Most complex attacks imitate ordinary HTTP traffic generated by botnets [2]. Attackers load scripts into infected botnet agents, which perform actions similar to those of ordinary users when they browse websites, but at high speed. The larger the botnet, the heavier the load it can produce on a target server. The destructive impact of a DDoS attack is that it significantly delays business processes [3, 4]. E-shops, news agencies, stockbrokers, banks, and many other types of businesses are very sensitive to stable continuous operation. Any, even short, interruption in the availability of their systems may lead to significant losses or even wide-scale disruption of the business.

In response to the threats described above, we realized the necessity of new technology for DDoS prevention. Our multilayer system implements both proactive preventive methods based on behavioral analysis, and threat intelligence, which in combination, provide proven attack prevention.

The main hypothesis of our research is that a highly effective system for DDoS protection in the cloud can be developed by taking into account the growing nature of the risk landscape.

Threat intelligence is a rapidly growing, though relatively young field of cyber security. Security vendors and independent researchers define this term as a complex process described by some common properties. We analyzed several definitions [5–9] and then combined them to form our own definition to highlight the most important properties and features of threat intelligence:

> *Threat Intelligence (TI) is a process to gather knowledge, aggregated from reliable sources, cross-correlated for accuracy. It must be timely, complete, assessed for relevancy, evaluated, and interpreted to create actionable data about known or unknown security threats that can be used to effectively respond to those threats*

The key benefits of using threat intelligence to prevent DDoS attacks are:

1. Protection of target websites from botnets (by implementing a botnet IP database and checking the incoming IPs against it), and DDoS attacks (by utilizing our five-layer system).
2. Decreasing the system load by blocking threats outside the target website perimeter (layer 2 determines the speed, layer 4 decreases the speed by the dispatcher and proxies).
3. Reducing system outages and cost of threats elimination and recovery (this is the general effect provided by our five-layer system: effective prevention of DDoS will eliminate system outages).
4. Automation of protection process from continuously growing threats (as mentioned above, we automate the prevention process by scripts used in our five-layer system).
5. Reduction of time needed to respond to new threats (because we use a proactive approach in our multilayer model, we are not required to wait until vendors identify new attack samples and update their signature bases).

Threat intelligence is a reliable modern technology to effectively protect against DDoS attacks and other threats taking into account the exponential growth of their complexity and intensity.

2 Background and Related Work

Cho et al. [10] proposed a DDoS prevention system based on the combination of a packet-filtering method with a double firewall. The first firewall analyzes the router path, whereas the second classifies data packets as being either normal or abnormal.

Botnets remain a highly destructive threat to cyber security. Graham et al. [11] attempted to detect botnet traffic within an abstracted virtualized infrastructure, such as that available from cloud service providers. They created an environment based on a Xen hypervisor, using Open vSwitch to export NetFlow Version 9. They obtained experimental evidence of how flow export is able to capture network traffic parameters for identifying the presence of a command-and-control botnet within a virtualized infrastructure. The conceptual framework they describe presents a non-intrusive detection approach for a botnet protection system for cloud service providers.

Karim et al. [12] reviewed methods of botnet detection and presented a method to classify botnet detection techniques. Their work highlights aspects pertaining to the analysis of these techniques with qualitative research design. The authors define possible future ways of improving the techniques of botnet detection and identify persistent research problems that remain open.

The evolution of DDoS attacks and their place in modern hybrid attacks and threats have been described in detail [13]. The nature of a DDoS attack, its effect on cloud computing, and problems that need to be considered while selecting defense mechanisms for DDoS were described in detail [14]. The authors' recommendation is to choose a functional, transpicuous, lightweight, and precise solution to prevent DDoS attacks, without any specific details.

The detection of DDoS attacks with the aid of correlation analysis formed the basis of research by Xiao et al. [15]. Their approach is based on a nearest-neighbors traffic classification with correlation analysis. It improves the classification accuracy by exploiting the correlation information of training data and reduces the overhead resulting from the density of training data.

Approaches to combatting both known and unknown DDoS attacks considering the real-time environment were described [16]. A method based on an artificial neural network (ANN) was used to detect attacks based on their specific patterns and characteristic features, thereby enabling these attacks to be distinguished from ordinary traffic.

3 Research Objectives and Methodology

3.1 Research Objectives

Let us define the aims and objectives of our research.

3.1.1 Defining Threat Intelligence and Its Scope

Our definition differs from other existing definitions, because in it we highlight all major properties of threat intelligence as a process to obtain knowledge. Most existing definitions describe threat intelligence as either a process or knowledge. However, it is neither knowledge nor simply a process; instead, it is a process to obtain actionable knowledge about both known and unknown threats. In our definition, we combine such mandatory properties as reliable sources, accuracy, completeness, relevancy, evaluation, interpretation, and being actionable.

3.1.2 Proposing an Innovative Method to Prevent DDoS Attacks in the Cloud Environment

Our method is different from existing methods, because we use a complex multi-layer system, which combines several techniques developed by us into an integrated system. In particular, we use our own enhanced method of IP traceback, own method of threat intelligence, own method of traffic dispatching, and own method of port hopping. The joint operation and interaction of these methods make our system unique and highly effective.

3.1.3 Introducing New Proactive Approach to Defend Threats Related to DDoS Attack

According to 2015 reports of major vendors [1], threats related to DDoS attacks have been increasing significantly. Moreover, almost no new types of attacks are invented. Instead, hackers improve old existing methods and add more power to them, for example, by using an amplification method. Their main aim is to exhaust system resources and overload the communication channels. That is why we can state that threats of DDoS attacks are critical today and can be expected to be of great importance in subsequent years as well.

3.1.4   Designing a Multilayer System for DDoS Prevention in the Cloud Using Threat Intelligence Techniques

Threat intelligence techniques are used by many existing solutions. In addition, IP traceback, port hopping, and many other techniques are used to prevent DDoS attacks. Yet, there is no effective solution in the world capable of really protecting against a DDoS attack. However, in the 21st century, many servers on the Internet can be shut down with a single command using SYN flooding. Other servers can be taken offline by DNS amplification requests or other very simple types of DDoS attacks. Our system is designed to provide a complex and integrated solution that uses the power of the best techniques, which were reinvented by us to solve existing problems and eliminate existing bottlenecks.

3.1.5   Introducing an Improved Method of IP Traceback

We named our method iDPM (improved Deterministic Packet Marking). It improves standard methods of the DPM type by using two octets of the options field, which allows us to store information about the route and IP address of the packet in full, without splitting it into two or more parts, as other methods do. Our method allows us to restore the full route on the victim's side and to protect it from packet loss by using the options field to repeat each IP address in two or more packets.

3.1.6   Introducing Our Own Simple and Effective Port-Hopping Method

Our port-hopping method uses unique formulas to calculate the port number. Moreover, we use a traffic dispatcher and proxy server(s) to add additional security to this method, because only the IP address of the dispatcher is visible from an external network. A malicious user would have to break both the dispatcher and proxy and would have to know the formulas to be able to spoof the port number and connect to the target website directly.

3.1.7   Experimental Confirmation of the Effectiveness of Our Method

We test and confirm the effectiveness of our method compared to other popular techniques, including IP traceback, port hopping, and entropy-based anomaly detection.

3.2   Methodology

The concept of our work is based on combining several protection methods and
adding a proactive approach with Threat Intelligence. We establish five protection
layers for all incoming traffic. The logic of these steps is detailed below. The logic
diagram is shown in Fig. 1.

3.2.1   Layer 1

At the first layer, we analyze the IP sources. If we find that a large amount of
anomalous traffic started from some range of IPs, we check these IP addresses to
determine whether they belong to Botnet IPs.

Packet-forwarding techniques such as NAT and encapsulation may be used on
the way of Internet traffic. Such techniques obfuscate the real origin of packets. We
analyzed the originating IP address using our traceback method, which we developed by analyzing existing IP traceback methods, selected the most appropriate
approach based on deterministic packet marking, and improved it.

Our method improves the approach followed previously [17–30] and represents
improved deterministic packet marking (iDPM) as having the best relation between
effectiveness and ease of implementation.

Usually, the field identification (16 bits), fragment offset (16 bits including
flags), DSCP (6 bits), ECN (2 bits), and even TTL (8 bits) are used in different
packet marking methods. As a result, the limitation of the size of these fields does
not allow the full IP to be stored in one packet; thus, it is fragmented. In our
method, we propose to use the Options field for our needs. It consists of 32 bits,

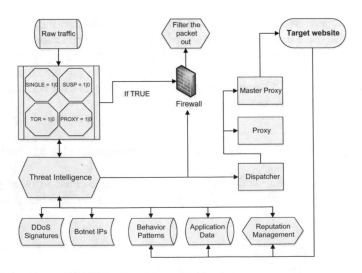

Fig. 1 Logic diagram of DDoS prevention method

which are necessary to store the full length of an IP address. Moreover, the IP header may have a variable number of options. If octet 20 is busy with real options, we may use the next octet. The maximum size of the header is 480 bits, of which only the first 160 bits (octets 0-16) are mandatory. Thus, we have 320 bits left for the Options field, which theoretically may be used to store up to 10 IPs. We use only one word (32 bits) from these 10 available to store one IP address without fragmentation.

Our aim is to trace the full route of the packets. For this purpose, we need to record the originating IP of the local computer (it may be in the format of a local network, e.g., 192.168.1.1) and the IPs of all routers through which the traffic passes. This approach allows us to trace the source IP even if NAT or proxies are used.

It is not a trivial task to detect traffic coming from botnets, because a good attack copies the user-agent of a genuine browser and imitates other signs of normal behavior. However, we can point out some initial indicators that would greatly help to reduce the power of an attack at the first two layers. These indicators include several variables (SUSP, SINGLE, PROXY, TOR) that we define and use in our method.

Traffic is considered suspicious (SUSP = TRUE) when a non-standard user-agent is detected. We allow search and stats bots, crawlers and validators as well as all standard browsers including mobile ones, but all others trigger this variable to TRUE, indicating the potential need for blocking. SINGLE = TRUE in case of a large number of requests from a single source. PROXY = TRUE indicates that the usage of a proxy server is detected. TOR = TRUE signifies the usage of Tor is detected by exit nodes or by checking the TorDNSEL value.

3.2.2 Layer 2

The second layer analyzes the speed of requests. If it is found that the rate at which inbound traffic is higher than a speed value, which is calculated below as a value of S (3), we can form blocking rules and pass them to the firewall. Otherwise, we simply pass the traffic to layer 3.

We capture the traffic by using any server tool that records the incoming traffic packets for 1 ms and counts the number of bits in the captured data. Then we multiply it by 1000 to obtain the number of bits per second.

The statistics of website visits may be taken from web analytics software such as Alexa, Google Analytics, and AWStats. We need unique visitors and the peak number of monthly visitors. Then we represent the numbers in the form of (1):

$$P = \begin{pmatrix} Range[a_1 - b_1] & peak_1 \\ Range[a_2 - b_2] & peak_2 \\ \cdots & \cdots \\ Range[a_n - b_n] & peak_n \end{pmatrix}, \tag{1}$$

where a range of monthly visitors (e.g., a_1-b_1) corresponds to the peak value of monthly visitors for the last three months for this range (e.g., $peak_1$). This information is useful to determine the possible attack speed threshold.

Assume the number of visitors for a day is U_i, where i is the day, d is the number of days (30), A is the number of visitors required to trigger an attack, then we multiply the corresponding number obtained from (1) by M to allow an excess number of visitors before we trigger an attack (the value of M is defined by experiments):

$$A = P\left[\left[\frac{\sum_{i=1}^{d} U_i}{d}\right]\right] * M. \tag{2}$$

The exact numbers in (1) may vary for different studies, but it does not affect the general formulas for A and S. This means the formulas we developed in this study will be universal for any other types of websites.

Once we know the number of visitors triggering the attack, we can calculate the speed of attack S:

$$S = \frac{A}{86400} * sizeof(packet) \tag{3}$$

where S is the rate at which we can consider traffic to be malicious in regard to DDoS attack activity.

3.2.3 Layer 3

Traditionally, Threat Intelligence is associated with the number of feeds received from many different sources. Special dedicated staff analyze these feeds for relevancy and all other properties. Although we also use feeds, our TI system is more complex in that it represents a combination of five modules. The TI architecture is shown in Fig. 2.

1. Behavioral Patterns. In our system, the criteria are specific to DDoS attacks and are provided by the target website we protect.
2. Application Data protects against ADDoS (Application DDoS) attacks.
3. Botnet IPs by third-party services.
4. DDoS signatures by third-party services.
5. Reputation Management (RM). We have identified our method of RM by calculating the reputation for each packet using the values of variables SUSP, SINGLE, PROXY, TOR, S, and the speed received from the previous layer.

We next define the formulas for attack detection A at a given time t–$A(t)$. Assume:

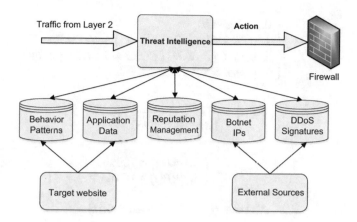

Fig. 2 Threat intelligence architecture

P pages per visit,
T time on site,
V new visitors,
B bounce rate,
R_N response time of a target website page,
N number of target website pages
RT reputation based on traffic variables
RP value of Reputation Management
A_1 attack detection for method 1 (behavior patterns),
A_2 attack detection for method 2 (application data)
A_3 attack detection for method 5 (reputation management)

Then—

$$A_1(t) = \begin{cases} 1, & |P_t \to 0 \\ 0, & |P_t \to \infty \\ 1, & |T_t \to 0 \\ 0, & |T_t \to \infty \\ 1, & V_t \to \infty \\ 0, & |V_t \to 0 \\ 1, & |B_t \to \infty \\ 0, & |B_t \to 0 \end{cases} \tag{4}$$

$$A_2(t) = \begin{cases} 0, & \left| \left(\frac{\sum_{i=1}^{N} R_{Ni}}{N} \right)_t \to 0 \\ 1, & \left| \left(\frac{\sum_{i=1}^{N} R_{Ni}}{N} \right)_t \to \infty \end{cases} \tag{5}$$

$$RT(t) = \begin{cases} 0, & |speed < S \\ 1, & |(PROXY = 1 \; OR \; TOR = 1) \; AND \; SUSP = 1 \\ 1, & |speed > S \; AND \; (PROXY = 1 \; OR \; TOR = 1) \\ 1, & |speed > S \; AND \; SUSP = 1 \end{cases} \tag{6}$$

$$RP(t) = \frac{RT(t)}{1/\left(\frac{\sum_{i=1}^{N} R_{Ni}}{N}\right)_t * S)} * 100\% \tag{7}$$

$$A_3(t) = \begin{cases} 0, & |RP(t) < 100\% \\ 1, & |RP(t) \geq 100\% \end{cases} \tag{8}$$

$$A(t) = A_1(t) \; AND \; A_2(t) \; AND \; A_3(t) \tag{9}$$

In the result, if $A(t) = 1$, we have an active attack at the given time, otherwise there is no attack.

3.2.4 Layer 4

The fourth layer dispatches the traffic to the proxy server to reduce traffic load, if necessary.

3.2.5 Layer 5

This is the last protection layer of our methodology and it strengthens our method by adding the port hopping technique. Our designed pseudo-random algorithm for changing port numbers resides in the fifth layer. This algorithm is known only to the proxy and target website.

$$Port(t) = (PRND0 \oplus t) \; mod \; 65535, \tag{10}$$

where PRND0 is the pseudo-random number generator, synchronized between the proxy and the target website, t is the current time and 65535 is the greatest possible port number.

4 Experiments

In the course of our study we conducted practical experiments to verify our assumptions and methods of attack detection and prevention. The aim of the experiment is to prove that our developed method to prevent DDoS attacks in the cloud is effective, accurate, and has strong advantages compared to other methods.

We tested source IP detection for different scenarios (with real and spoofed source IP) for layer 1, and then calculated the speed for layer 2 for low speed, normal speed, and high speed attacks. Figure 3 displays the charts for different speed, and Table 1 lists the values for the corresponding speed.

Then we calculated the values of A1 for behavioral patterns, A2 for application data, A3 for reputation management, and the resulting A indicating the attack at layer 3.

After that we divided the speed by Dispatcher at layer 4, as shown in Fig. 4. Then we checked the generation of port numbers for the port hopping method at layer 5. We ran the script generating the port numbers using our method, and we confirmed the numbers were random and different each time.

Lastly, we ran experiments using the overall method that resulted in the generation of firewall blocking rules. Table 2 contains the results of launching attacks from 1, 2, and 5 of our 5 VM clients.

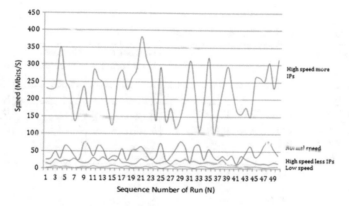

Fig. 3 Source IP detection at different traffic speeds

Table 1 Values of speed

Seq no. of run	Low speed	Normal speed	High speed (fewer IPs)	High speed (more IPs)
1	1,5	25,5	15,3	231,6
2	2,8	28,3	12,8	227,8
3	5,7	49,1	25,1	235,2
4	3,2	28,6	19,5	351,2
5	5,7	65,1	23,2	256,3
6	3,2	56,4	21,3	223,6
7	0,9	35,6	28,3	136,6
8	2,8	25,3	18,4	187,4
9	4,3	74,2	15,2	236,9
10	3,2	64,2	17,9	167,4

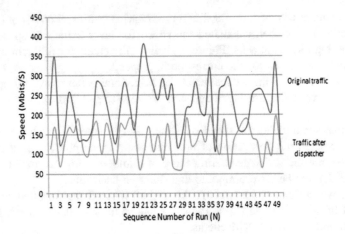

Fig. 4 Speed before and after processing by the dispatcher at layer 4

Table 2 Experimental results for overall method

IPs of attack	Number of IPs detected	Number of IPs blocked	False positive rate, %
VM6 192.168.0.141	1	1	0
VM6 192.168.0.141 VM7 192.168.0.138	2	2	0
VM6 192.168.0.141 VM7 192.168.0.138 VM8 192.168.0.139 VM9 192.168.0.143 VM10 192.168.0.144	5	5	0

Table 3 Comparison of methods

	Our method	Method 1[a] [31]	Method 2[b] [32, 33]	Method 3[c] [34]
Usage of multiple technologies	yes	no	no	no
Number of technologies used	6	1	1	1
Requiring of user's actions	no	yes	no	no
False positive rate	0%	50%	50%	30%
Average Response time (ms) of target on low load	0.346	0.564	0.287	0.334
Average Response time (ms) of target on high load	0.513	0.974	0.816	0.806
Dependence on signatures	no	no	no	no
Ability to detect unknown threats	yes	yes	yes	yes
Practical implementation	yes	yes/no	yes	yes
Overall System	excellent	good	good	good

[a]IP traceback
[b]Port hopping
[c]Entropy-based anomaly detection

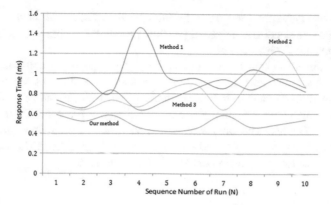

Fig. 5 Response time for different methods

Thus, the result is confirmed, and the attack is successfully detected and blocked. Furthermore, we compared our method to other popular methods used for DDoS protection. These results are presented in Table 3 and shown in Fig. 5.

Thus, all experiments were successful and we met the expected results. Our main hypothesis is confirmed, and the attacks we successfully detected and blocked.

In addition, we compared our method to other popular methods used for DDoS protection.

For example, Fig. 5 below shows the response time in milliseconds for different methods using a high traffic load:

5 Conclusions

The hypothesis of our research is that a highly effective DDoS protection system for the cloud can be developed taking into account the growing nature of the risk landscape. Our study and the set of experiments we carried out showed that existing methods have shortcomings and they could not positively confirm this hypothesis. At the same time, our proposed method introduced enhancements to existing techniques such as IP traceback, port hopping, and reputation management. Moreover, we introduced a completely new definition and methods for threat intelligence and the results of our experiments confirmed that the definition occupies a central part of our protection method and allows complex DDoS attacks to be prevented proactively without human intervention. Thus, we confirmed our main hypothesis and we showed that our method produces 0% false positives, a minimal response time on target with and without load, the ability to detect unknown threats, and a high level of practical implementation.

We would like to suggest possible follow-up studies in response to our research, related to different combinations of DDoS protection techniques in one complex multilayer system such as ours.

There are many possible ways to incorporate existing or newly developed methods into one system, a variety of possible protocols to achieve their interaction, and their potential enhancements.

Modern threats dictate the overestimation of the potential consequences of a successful attack and require us to always be a step ahead of malicious attackers. This in turn requires new complex methods to be used instead of a single technology. No standalone technology today would stop a powerful DDoS reflection attack over 500 Gb/s. The only way to be victorious over cybercriminals is to combine the efforts of scientists and security vendors to produce all-in-one solutions that would proactively mitigate attacks of any given type.

References

1. Akamai, State of the Internet Report (2015).
2. Wang, A., Mohaisen, A., Chang, W., Chen, S.: Delving into internet DDoS attacks by botnets: characterization and analysis. In: 45th Annual IEEE/IFIP International Conference Dependable Systems and Networks (DSN), 379–390 (2015).
3. Arbor Networks. Worldwide Infrastructure Security Report, DDoS Threat Landscape. APNIC Conference (2016).
4. Riverhead Networks. DDoS Mitigation: Maintaining Business Continuity in the Face of Malicious Attacks, Cupertino: Riverhead, Cisco (2004).
5. Friedman, J., Bouchard, M.: Definitive Guide to Cyber Threat Intelligence, CyberEdge Press (2015).
6. Cyber threat intelligence - how to get ahead of cybercrime, Ernst & Young Global Limited (2014).
7. Chismon, D., Ruks, M.: Threat Intelligence: Collecting, Analysing, Evaluating. MWR InfoSecurity Ltd (2015).
8. Farnham, G., Leune, K.: Tools and standards for cyber threat intelligence projects, SANS Institute (2013).
9. McMillan, R.: Definition: Threat Intelligence. Gartner, 2013.
10. Cho, J.H., Shin, J.Y., Lee, H., Kim, J.M., Lee, G.: DDoS Prevention System Using Multi-Filtering Method (2015).
11. Graham, M., Winckles, A., Sanchez-Velazquez, E.: Botnet detection within cloud service provider networks using flow protocols. In: IEEE 13th International Conference on Industrial Informatics (INDIN), 1614–1619 (2015).
12. Karim, A., Salleh, R.B., Shiraz, M., Shah, S.A.A., Awan, I., Anuar, N.B.: Botnet detection techniques: review, future trends, and issues. Journal of Zhejiang University SCIENCE C 15, 943–983 (2014).
13. Mansfield-Devine, S.: The evolution of DDoS. Computer Fraud & Security 2014, 15–20 (2014).
14. Deshmukh, R.V., Devadkar, K.K.: Understanding DDoS Attack & its Effect in Cloud Environment. Procedia Computer Science 49, 202–210 (2015).
15. Xiao, P., Qu, W., Qi, H., Li, Z.: Detecting DDoS attacks against data center with correlation analysis. Computer Communications 67, 66–74 (2015).
16. Saied, A., Overill, R.E., Radzik, T.: Detection of known and unknown DDoS attacks using Artificial Neural Networks. Neurocomputing 172, 385–393 (2016).
17. Ramesh, S., Pichumani, S., Chakravarthy, V.: Improving the Efficiency of IP Traceback at the DoS Victim. http://www.cs.utah.edu/~sramesh/attachments/ip_traceback.pdf.

18. Saurabh, S., Sairam, A.S.: Increasing Accuracy and Reliability of IP Traceback for DDoS Attack Using Completion Condition. Int. J. Network Security **18**, 224–234 (2016).
19. Li, J., Sung, M., Xu, J., Li, L.: Large-scale IP traceback in high-speed Internet: Practical techniques and theoretical foundation. In: Proceedings of the IEEE Symposium on Security and Privacy, 2004. 115–129 (2004).
20. Gong, C., Sarac, K.: IP traceback based on packet marking and logging. In: IEEE Conference on Communications (ICC). **2**, 1043–1047 (2005).
21. Foroushani, V.A., Zincir-Heywood, A.N.: Deterministic and authenticated flow marking for IP traceback. In: IEEE 27th International Conference on Advanced Information Networking and Applications (AINA), 397–404 (2013).
22. Yan, D., Wang, Y., Su, S., Yang, F.: A precise and practical IP traceback technique based on packet marking and logging. J. Inf. Sci. Eng. **28**, 453–470 (2012).
23. Aghaei-Foroushani, V., Zincir-Heywood, A.N.: On evaluating IP traceback schemes: a practical perspective. In IEEE Security and Privacy Workshops (SPW), 127–134 (2013).
24. Sung, M., Xu, J. IP traceback-based intelligent packet filtering: a novel technique for defending against Internet DDoS attacks. IEEE Trans. Parallel Distrib. Syst. **14**, 861–872 (2003).
25. Park, K., Lee, H.: On the effectiveness of probabilistic packet marking for IP traceback under denial of service attack. In *INFOCOM 2001*. Twentieth Annual Joint Conference of the IEEE Computer and Communications Societies. IEEE Proceedings **1**, 338–347 (2001).
26. Song, D.X., Perrig, A.: Advanced and authenticated marking schemes for IP traceback. In: INFOCOM 2001. Twentieth Annual Joint Conference of the IEEE Computer and Communications Societies. IEEE Proceedings, **2**, 878–886 (2001).
27. Parashar, A. Radhakrishnan, R.: Improved deterministic packet marking algorithm for IPv6 traceback,. In: International Conference on Electronics and Communication Systems (ICECS), 1–4 (2014).
28. Amin, S.O., Hong, C.S.: On IPv6 Traceback. In: The 8th International Conference on Advanced Communication Technology, ICACT 2006. **3**, 2139–2143 (2006).
29. Amin, S.O., Kang, M.S., Hong, C.S.: A lightweight IP traceback mechanism on IPv6. In: Emerging Directions in Embedded and Ubiquitous Computing, Amin, S.O., Kang, M.S., Hong, S.C. (Eds.) Springer, Berlin Heidelberg (2006).
30. Kim, R.H., Jang, J.H., Youm, H.Y.: An Efficient IP Traceback mechanism for the NGN based on IPv6 Protocol, IITA'09 (2009).
31. Savage, S., Wetherall, D., Karlin, A., Anderson, T.: Practical network support for IP traceback. In: ACM SIGCOMM Computer Communication Review, **30**, 295–306 (2000).
32. Shi, F.: U.S. Patent No. 8,434,140. Washington, DC: U.S. Patent and Trademark Office (2013).
33. Morris, C.C, Burch, L.L., Robinson, D.T.: U.S. Patent No. 8,301,789. Washington, DC: U.S. Patent and Trademark Office (2012).
34. Source code of the entropy-based network traffic anomaly detector. Retrieved May 26, 2016, from https://github.com/anacristina/entropy-based-anomaly-detector.

Practical Uses of Memory Storage Extension

Shuichi Oikawa

Abstract Memory storage technologies are emerging. Battery backed NV-DIMMs are becoming widely available, and the general availability of next generation non-volatile (NV) memory is also considered to be close. While memory storage provides much higher performance than the current block storage devices, such as SSDs and HDDs, its capacity tends to be limited. Such a limitation makes it difficult to adapt memory storage for mass storage; thus, its uses have been limited. Memory storage extension, which we call MSX, is an operating system mechanism that has a file system directly access memory storage and also that virtually extends the capacity of memory storage to that of block storage; thus, MSX effectively utilizes the high performance of memory storage by having a file system directly access memory storage through the synchronous memory interface, and also provides the large capacity by employing block storage as backing storage. MSX was implemented in the Linux operating system kernel. This paper discusses the several practical uses of MSX in cloud computing and also in the fundamental operating system architecture.

1 Introduction

Memory storage technologies are emerging. Battery backed NV-DIMMs are becoming widely available, and the general availability of next generation non-volatile (NV) memory devices, such as MRAM, PCM (phase change memory), and ReRAM, is also considered to be close. Memory storage are byte addressable; thus, it is called memory storage. It provides persistency since it is storage. It also provides much higher performance than block storage. Since their byte addressability enables them to be accessed as memory, we call them *memory storage*. While its performance is much higher than block storage, its capacity tends to be limited. Its capacity limitation narrows the areas where memory storage can fit.

S. Oikawa (✉)
Faculty of Engineering, Information and Systems, University of Tsukuba,
1-1-1 Tennodai, Tsukuba, Ibaraki, Japan
e-mail: shui@cs.tsukuba.ac.jp

© Springer International Publishing AG 2017
R. Lee (ed.), *Applied Computing and Information Technology*,
Studies in Computational Intelligence 695,
DOI 10.1007/978-3-319-51472-7_10

Memory Storage Extension, which we call MSX, is an operating system mechanism that has a file system directly access memory storage and also that virtually extends the capacity of memory storage to that of block storage [10]. A technique to combine block storage with another faster block storage, which is typically an SSD, for higher access performance is broadly recognized to be useful and is widely utilized [5, 6, 12]. Faster block storage is used as cache and stores frequently accessed data. Such a use of SSDs as cache improves the overall time to access data. Its open source implementation is available [3], and also the current Linux kernel includes several implementations, such as dm-cache and bcache. The existing technique implements its mechanism as a software layer in the block device driver framework, that combines two block storage devices. Memory storage can fit in such a software layer by having it emulate block storage. The use of the block device driver framework, however, sacrifices the performance advantage of memory storage for the compatibility with block storage.

MSX takes a very much different approach from the combined devices described above in the following three ways. First, MSX combines memory storage and block storage, and it make use of the byte addressable feature of memory storage by exposing it rather than sacrificing it under the interface of block storage. While MSX is implemented in the Linux kernel, it does not use the device mapper mechanism because of its overheads and its use of the block storage interface. Second, MSX makes use of synchronous access when data is available on memory storage. When data needs to be retrieved from block storage, asynchronous access is used. Such adaptive use of synchronous and asynchronous access significantly reduces the access cost in total. MSX extended the DAX interface, and implements its own functions for the direct and synchronous access to memory storage. Third, MSX is an architecture that has a file system directly access memory storage. A file system does not employ a device driver to access memory storage since processors can access memory storage directly, and such direct access enables a file system to use its knowledge for more efficient access. MSX's involvement of a file system takes the memory management one step further towards the single-level store [2, 13].

MSX was implemented in the Linux kernel. It was implemented by extending the DAX interface, which was introduced in Linux 4.0 as an interface to support memory storage [1]. DAX can access only memory storage, so that its capacity is limited to that of memory storage. MSX extends the capacity of memory storage by employing a block storage device as a backing store; thus, the capacity enabled by MSX is that of block storage. Thus, the implementation of MSX generalized the interface to memory storage by providing the abstraction of virtualized memory storage.

This paper discusses the four practical uses of MSX. First, two of MSX's uses in cloud computing are described. MSX accelerates the file access performance of Hadoop Distributed File System (HDFS) approximately 4.12x faster, and MSX enables the quick system rejuvenation that is 10x faster than the existing method. Next, two applications of MSX in the fundamental OS architecture are described. MSX can accelerate the journaling mechanism by its direct access to memory storage. And, MSX takes the memory management one step further towards the single-level store by involving memory storage in the hierarchy of memory and storage.

The rest of this paper is organized as follows. Section 2 describes the background of the work. Section 3 describes the detailed design and implementation of MSX. Section 4 discusses the practical uses of MSX. Section 5 describes the related work. Section 6 summarizes the paper.

2   Background

It is important to understand the high software overhead to access block storage since long access times of block storage have hidden such overhead. Short access times of memory storage, however, change the rule. Therefore, this section describes the background of this work. It includes the overview of the block device driver layer of the operating system (OS) kernel and the existing method to combine block storage devices [8].

2.1   Block Device Driver Layer and Its Execution Overhead

The current storage devices, such as SSDs and HDDs, are block devices, and they are not byte addressable, thus, CPUs cannot directly access the data on these devices. A certain size of data, which is typically multiples of 512 byte, needs to be transferred between memory and a block device for CPUs to access the data on the device. Such a unit to transfer data is called a block.

The OS kernel employs a file system to store data in a block device. A file system is constructed on a block device, and files are stored in it. In order to read the data in a file, the data first needs to be read from a block device to memory. If the data on memory was modified, it is written back to a block device. A memory region used to store the data of a block device is called a page cache. Therefore, CPUs access a page cache on behalf of a block device. Figure 1 depicts the hierarchy of CPUs, a file system, page cache, a block device driver, and a block device as the existing architecture to interact with block devices.

Since HDDs are orders of magnitude slower than memory to access data on them, various techniques were devised to amortize the slow access time. The asynchronous access command processing is one of commonly used techniques. Its basic idea is that a CPU executes another process while a device processes a command. In Fig. 1, the I/O request queueing mechanism in a block device driver provides the asynchronous access command processing. Figure 2 depicts how it works. Process 1 issues a system call to access data on a block device. The kernel processes the system call and issues an access command to the corresponding device. The kernel then looks for the next process to execute and perform context switching to Process 2. Meanwhile, the device processes the command, and sends an interrupt to notify its completion.

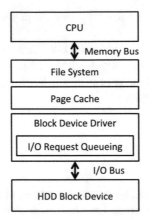

Fig. 1 The existing operating system architecture to interact with block devices

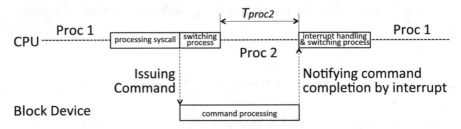

Fig. 2 The asynchronous access command processing and process context switches

The kernel handles the interrupt, processes command completion, and performs context switching back to Process 1. T_{proc2} is a time left for Process 2 to run. Because HDDs are slow and thus their command processing time is long, T_{proc2} is long enough for Process 2 to proceed its execution.

The I/O request queueing mechanism that implements the asynchronous access command processing has been a right choice for block devices. While it poses high processing cost, the cost pays off by creating additional processing times made available for other processes. Such a justification, which stands for the I/O request queueing mechanism and the asynchronous access command processing, is no longer true when storage becomes much faster. Figure 3 shows the comparison of the execution costs for read with and without I/O request queueing on a ramdisk device. The measurements were performed on the Linux kernel, and they counted the number of executed instructions. Because the storage device used for the measurements is a ramdisk, its access cost is basically negligible; thus, the results reflect purely the software costs to access storage. The results unveil the significant overhead of the I/O request queueing mechanism that has been hidden in the long access latency of block storage devices. Therefore, high performance memory storage and slower block storage have to be managed differently.

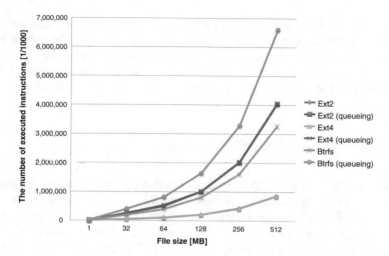

Fig. 3 Comparison of the execution costs for read with and without I/O request queueing on a ramdisk device

2.2 The Existing Method to Combine Block Storage Devices and Its Problems to Be Used with Memory Storage

The existing method combines block storage with another faster block storage, which is typically an SSD, for higher access performance [5, 6, 12]. It utilizes faster block storage as cache and store frequently accessed data in it in order to improve the average time to access data. Its open source implementation is widely available [3], and the current Linux kernel includes several implementations, such as dm-cache and bcache.

The implementations of the existing method in the Linux kernel employ the device mapper as the software layer to constitutes a single storage device. The device mapper provides the mechanism to transfer (or *map*) access requests for the consti-tuted device to underlying devices. The policy part defines how it maps requests. There can be multiple policy implementations, and some of them combines block storage with faster storage as cache. When an SSD is used as cache storage by com-bining it with a HDD, it is straightforward that the combined storage provides the block storage interface and is accessed asynchronously since both of them are block storage. As its extension, it is possible for memory storage to emulate block storage and to have the device mapper to combine block storage with memory storage.

The use of the device mapper requires memory storage to emulate block storage since the device mapper can interact only with the block storage interface. While such emulation enables the reuse of the device mapper, it causes significant software overhead. The device mapper is basically a block device driver; thus, it receives access requests from the upper generic block device driver framework. It then trans-fers the received requests to another block storage device. The transferred requests are processed again by the generic block device driver framework, and the target block storage device receives them [15]. Therefore, processing in the generic block device driver framework occurs multiple times, and such multiple times processing causes a software overhead that can be hidden in the long access latency of block storage devices but can be significant for high performance memory storage.

3   MSX: Memory Storage Extension

This section describes MSX, Memory Storage Extension, which enables file systems to directly and synchronously access memory storage without employing a device driver [10]. MSX was developed with the two goals, (1) introduction of memory storage in the file system layer to provide the capacity of block storage, and (2) provision of efficient use of memory storage by direct and synchronous access to memory storage. In order to achieve these goals, the MSX interface was designed in a way that the file system layer takes control of memory storage. They led a design that the file system layer access memory storage directly. This design is distinguished for efficiently utilizing memory storage since its management can make use of the rich information of files. If memory storage is managed indirectly through the block device interface, such knowledge cannot be used, and its management becomes inefficient.

The overview architecture of MSX is depicted in Fig. 4. MSX resides within the file system layer, and controls the access to both memory storage and block storage. It accesses memory storage directly and synchronously while it accesses block storage indirectly and asynchronously through a block storage device driver. The direct and synchronous path to memory storage is shown by (a) of the figure, and the indirect and asynchronously path to block storage is shown by (c) of the figure. The primary storage remains block storage, and memory storage work as the cache of block storage. MSX is responsible to manage the consistency and transparency of data by transferring data between memory storage and block storage appropriately, which shown by (b) of the figure.

There are several advantages of having the file system layer manage memory storage directly. One is that it enables the MSX interface to utilize the rich information

Fig. 4   The design overview of MSX, which enables file systems to directly access memory storage without interposing a device driver

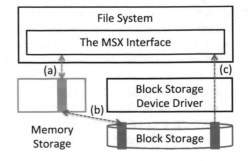

of files. Block storage device drivers receive only block numbers. While a sequence of block numbers constitutes a file, there is no way for block storage device drivers to know what are the following block numbers, what kind of a file it is, how often it is accessed, and so on. They are the information only available in file systems. Such information can be used to enable intelligent operations that improve the efficiency of cache management on memory storage in order to accelerate access performance. Sharing memory storage among multiple storage devices is also an advantage of MSX. While it is common to use multiple block storage devices, it is inefficient to dedicate a portion of memory storage to a specific block storage device. MSX can efficiently utilize memory storage to cache data of multiple block storage devices.

MSX disables the page cache. When the MSX interface is enabled, a file system assumes that data is or will be available in memory storage. Since memory storage is byte addressable and processors can access on it directly, there is no need to use the page cache. Disabling the page cache is implemented since the MSX is implemented by extending the DAX mechanism [1]. DAX was introduced in Linux 4.0 as an interface to support memory storage. Since memory storage can be accessed directly, there is no need of the page cache. DAX can access only memory storage while MSX extends the capacity of memory storage by employing a block storage device as a backing store. Thus, the implementation of MSX generalized the interface to memory storage by providing the abstraction of virtualized memory storage.

4 Practical Uses of MSX

This section describes the practical uses of MSX. First, two of MSX's uses in cloud computing are described, high performance file access and quick system rejuvenation. Next, MSX's applications in the fundamental OS architecture are described, file system journaling acceleration and single-level store. While the measurement results that are shown below include the results measured by using our previous implementation [9], MSX is its refactored implementation giving more control to file systems and thus provides the same or better performance.

4.1 High Performance File Access for Cloud Computing

Computing is an act of transforming data to another; thus, file access performance is an important factor to achieve faster computation. Of course, cloud computing is not an exception since it is widely utilized as a basis of big data computing.

Hadoop Distributed File System (HDFS) was first implemented as a file service for Hadoop [14], it is widely used by the other data processing systems. Such examples include HBase, Storm, SPARK, and Tez. HDFS provides its clients with high performance file access by placing the clients' computation near data and accessing files through sequential I/O. When the clients' computation is performed on the

same server node where the files they access reside, it can access files the most efficiently. HDFS provides the short circuit read feature in order to accelerate read performance for local clients. It enables clients executing locally to directly read data
files by receiving their file descriptors from the corresponding data service nodes;
thus, there is no need for clients to communicate with data service nodes through
RPCs.

The short circuit read feature best suits MSX since memory storage can work
as a large read-ahead buffer. Since MSX is implemented in the file system layer, it
can make use of the knowledge available in the layer. Read-ahead is an example of
such use of the knowledge. MSX can read ahead the exact data of a file which is
currently read, and such exact read-ahead is not possible in the device driver layer.
Although not implemented yet, it is possible for HDFS's client to read data directly
from memory storage without copying data to the buffer allocated in the client by
having MSX allocate a specific region on memory storage for HDFS. MSX provides
the dedicated buffer for HDFS on memory storage, and maps the buffer in the address
space of the client. In this way, since MSX feeds data as HDFS consumes it, MSX
reads data from block storage in parallel with the client's computation.

We measured the HDFS read performance with the short circuit read feature
enabled by using the Hadoop TestDFSIO [8]. While TestDFSIO is a simple benchmark program, it is written in the Java language within the same framework as the
other Hadoop applications; thus, it consults the service nodes of HDFS for its execution. Figure 5 shows the result. For reading from 100MB to 1GB file sizes, it performs approximately 3.66x to 4.40x faster than the SSD. On average, MSX makes
TestDFSIO perform 4.12x faster than the SSD.

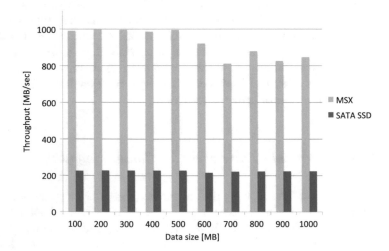

Fig. 5   Read performance of Hadoop HDFS Test DFSIO with short circuit read

4.2 Quick System Rejuvenation for Higher Reliability

MSX enables the zero copy checkpoint and restart, and one of its application is the quick system rejuvenation. The system reliability is becoming important more and more in cloud computing as it is utilized more widely as utility. The rejuvenation is a technique to increase system reliability, and the quick system rejuvenation enables it at mostly the constant cost. The constant cost of the quick system rejuvenation is a huge advantage in comparison with the existing rejuvenation method, of which cost heavily depends of the system memory size.

The quick system rejuvenation combines the image based kernel restart and the zero copy checkpoint/restart [7]. The image based kernel restart is a technique to shorten the time to reboot the kernel by restoring the kernel image. The zero copy checkpoint/restart is a technique specifically enabled by MSX. The checkpoint of a user process saves the process memory pages in a file, and the restart restores them from a file. Since MSX can make the memory pages a part of a file instantly, there is no (zero) copy involved in creating a checkpoint file. Thus, the zero copy checkpoint/restart significantly accelerates the process of checkpoint/restart of user processes.

Figure 6 shows the comparison of system rejuvenation costs that cumulate the costs of kernel restart and the zero copy checkpoint/restart of user processes. The zero copy checkpoint is 8x to 10x faster than the exiting checkpoint, and the zero copy restart is 28x to 36x faster the existing one. The significant advantage of the zero copy checkpoint/restart is that its cost does not go up rapidly as the memory size of a user process increases. While the memory size of a user process significantly impacts the existing checkpoint/restart, it impacts much less the zero copy checkpoint/restart. The image based kernel restart is 10x faster than the ordinal

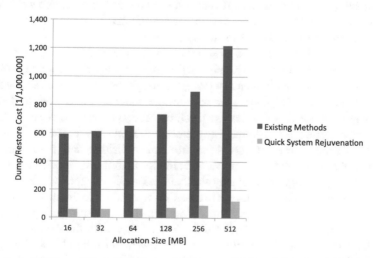

Fig. 6 System rejuvenation costs that cumulate the costs of kernel restart and zero copy checkpoint/restart of user processes

kernel reboot. The quick system rejuvenation, which combines the image based kernel restart and the zero copy checkpoint/restart, is 10x faster than the combination of the ordinal kernel reboot and the existing checkpoint/restart.

4.3 Journaling Acceleration

MSX can accelerate the journaling mechanism of file systems by its direct access memory storage. The journaling is a mechanism to guarantee the consistency of written data. It is known as write ahead logging (WAL) for database management systems. The journaling writes data twice, first in the journal and second in the destination place. If writing in the journal is aborted by a failure, the data the destination place remains consistent. If writing in the destination place is aborted by a failure, the correct data in the journal can repair it. Writing in the journal is also called logging. Logging is performed ahead of writing; thus, it is called write ahead logging. There can be various implementations of the journaling. For example, the Ext4 file system of the Linux kernel provides the three journal modes, journal, ordered, writeback. The journal mode first logs metadata and data into the journal, and then write them in place. The ordered mode first writes data, second logs metadata, and then write it in place. The writeback mode does not guarantee the order of metadata and data writes as the other two do. The ordered mode is used by default.

There are the two significant cost of the journaling. One is a latency to complete writing, and the other is the amount of data to be written. Logging must be completed and it must ensure the log becomes persistent before writing in the destination place. In other words, writing in the destination place must wait for the completion of logging. Each log tends to be small data, and writing small data in block storage is a typical inefficient operation; thus, it causes a long latency to complete writing. The journaling must write data twice, first in the journal and second in the destination place; thus, the cost of writing data simply doubles.

MSX can significantly reduce these two costs by the two features, synchronous write and write redirection. Synchronous write solves the latency problem. MSX can synchronously write data in memory storage through the memory access interface. It is possible since MSX places memory storage under its direct control and does not employ a device driver to access it. Direct and synchronous write of the log data in memory storage can dramatically shorten the time to wait for the completion of logging. Write redirection solves the problem of writing data twice. Since the capacity of memory storage is limited and smaller than that of block storage, dirty data in memory storage is eventually written back to block storage. MSX controls write back of data in memory storage to block storage. Write redirection enables data in memory storage to be written at multiple places in block storage. By using write redirection, it is possible to associated the logged data with the destination place. When the data associated by write redirection is written back to block storage, it is written back both in the journal and the destination place; thus, there is no need to write data twice.

4.4   Single-Level Store

MSX can be considered as an implementation of the single-level store [2]. The single-level store is the memory mapping model that objects represent data storage and that objects are accessed through the memory interface [13]. From the point of view of the single-level store, memory and disk storage are distinct parts of computer systems since memory is byte addressable while disk storage is block addressable; thus, processors cannot access disk storage directly, and data on disk storage must be brought to memory in order for processors to access it. Files are the abstraction of disk storage, and file systems manage disk storage to provide storage spaces with users as files. Data in files is access through the file API, which is designed to deal with block addressable disk storage.

Modern operating system design blurs the distinction between memory and disk storage by making memory to be cache of disk storage [11]. Memory was used to buffer I/O to disk storage in order to accelerate I/O by merging small requests to larger ones even before the above work. Memory used for such buffering is, however, separated from memory allocated for processes. In order for a process to access data on disk storage, data on disk storage is first transferred to buffer memory, and then the data in the buffer memory is copied to the process memory; thus, the buffer memory and the process memory are distinct from each other, and the buffer memory could not be used as the process memory. Making memory to be cache of disk storage removes such distinction and integrates the management of buffer memory and process memory. The integration enables cache memory of disk storage to be used as both buffer memory and process memory.

MSX takes the memory management one step further towards the single-level store by involving non-volatile memory storage in the hierarchy of memory and storage. Processors can access data on memory storage directly since memory storage is byte addressable. When all of storage is memory storage, processors can access all data objects on memory storage; thus, memory storage realizes the single-level store. In this way, a file system on memory storage serves a protection and name service. A file system servers a protection service since it allows only users, who have the appropriate rights, to access the data. A file system serves a name service since it provides the location of data from the name of the data, which is a file path name of the file system. Once a user process obtains the location of data on memory storage, it is directly accessed through the memory interface. The written data on memory storage is persistent unlike the data written on DRAM cache memory, which is not persistent until it is written to persistent block storage. MSX employs block storage to extend the capacity of memory storage. Such extension of the capacity is encapsulated under the interface of a file system, and only the view of memory storage can be exported to user processes. Therefore, MSX can work as the single-level storage.

5   Related Work

Combining block storage with another for higher access performance is a technique that existed before SSDs become widely available and popular. DCD [4] proposed a method that first stores data sequentially in cache storage, so that it can make use of higher performance sequential access; thus, the write performance can be improved. As SSDs emerged, their performance, which is much higher than HDDs, stimulated the research and development of various techniques that cache data in SSDs [3, 5, 6, 12]. SSDs and HDDs are block storage, and all of the above existing methods combine block storage devices with the other ones, and the combined devices provide the interface of block storage.

The device mapper of the Linux kernel is the generic software layer that combines multiple storage devices to constitutes a single storage device. The device mapper allows multiple policies; thus, using a storage device as cache is one of its usages. When the device mapper combines memory storage with block storage, the resulting device is block storage. Since it combines multiple block storage devices, it requires a memory storage device driver that provide the block storage interface. The use of the block storage interface sacrifices the nature of memory storage, and becomes the source of the significant software overhead since the generic block device driver framework is invoked multiple times to process a single request [15].

MSX is very much different from the combined devices described above in the following three ways. First, MSX combines memory storage and block storage, and it make use of the byte addressable feature of memory storage by exposing it rather than sacrificing it under the interface of block storage. While MSX was implemented in the Linux kernel, it does not use the device mapper mechanism because of its overheads and its use of the block storage interface. Second, MSX makes use of synchronous access when data is available on memory storage. When data needs to be retrieved from block storage, asynchronous access is used. Such adaptive use of synchronous and asynchronous access significantly reduces the access cost in total. MSX extended the DAX interface, and implements its own functions for the direct and synchronous access to memory storage. Third, MSX is an architecture that have a file system directly access memory storage. A file system does not employ a device driver to access memory storage since processors can access memory storage directly, and such direct access enables a file system to use its knowledge for more efficient access. MSX's involvement of a file system takes the memory management one step further towards the single-level store.

6   Summary

Memory storage technologies are emerging. Battery backed NV-DIMMs are becoming widely available, and the general availability of next generation non-volatile (NV) memory is also considered to be close. While memory storage provides much higher performance than the current block storage devices, such as SSDs and HDDs, its capacities tend to be limited. Such limitation makes it difficult to adapt memory storage for mass storage; thus, its uses have been limited. MSX is an architecture that have a file system directly access memory storage and also that virtually extends the capacity of memory storage to that of block storage; thus, MSX effectively utilizes the high performance of memory storage by having a file system directly access memory storage through the synchronous memory interface, and also provides the large capacity by employing block storage as backing storage. MSX was implemented in the Linux kernel.

This paper discussed the four practical uses of MSX. First, two of MSX's uses in cloud computing were described, high performance file access and quick system rejuvenation. Next, MSX's applications in the fundamental OS architecture were described, file system journaling acceleration and single-level store. These practical uses ensure that MSX can be a basis of not only specific uses but broader applications.

References

1. Supporting filesystems in persistent memory (2014). http://lwn.net/Articles/610174/
2. Clark, B.E., Corrigan, M.J.: Application system/400 performance characteristics. IBM Systems Journal 28(3), 407–423 (1989). doi:10.1147/sj.283.0407
3. Facebook: Flashcache (2014). https://github.com/facebook/flashcache
4. Hu, Y., Yang, Q.: Dcd – disk caching disk: A new approach for boosting i/o performance. In: Proceedings of the 23rd Annual International Symposium on Computer Architecture, pp. 169–178 (1996). doi:10.1109/ISCA.1996.10021
5. Kgil, T., Mudge, T.: Flashcache: A nand flash memory file cache for low power web servers. In: Proceedings of the 2006 International Conference on Compilers, Architecture and Synthesis for Embedded Systems, CASES '06, pp. 103–112. ACM, New York, NY, USA (2006). doi:10.1145/1176760.1176774
6. Koller, R., Marmol, L., Rangaswami, R., Sundararaman, S., Talagala, N., Zhao, M.: Write policies for host-side flash caches. In: Proceedings of the 11th USENIX Conference on File and Storage Technologies, FAST'13, pp. 45–58. USENIX Association, Berkeley, CA, USA (2013). http://dl.acm.org/citation.cfm?id=2591272.2591278
7. Oikawa, S.: Independent kernel/process checkpointing on non-volatile main memory for quick kernel rejuvenation. In: Proceedings of International Conference on Architecture of Computing Systems, ARCS'14, pp. 234–245. Springer (2014)
8. Oikawa, S.: Accelerating storage access by combining block storage with memory storage. In: Proceedings of 14th IEEE/ACIS International Conference on Computer and Information Science, ICIS 2015, pp. 449–454. IEEE Computer Society (2015)
9. Oikawa, S.: Exposing non-volatile memory cache for adaptive storage access. In: Proceedings of 30th ACM Symposium On Applied Computing, SAC 2015, pp. 2021–2026. ACM (2015)
10. Oikawa, S.: Msx: Memory storage extension for linux 4.0 and beyond. In: Proceedings of ACM Conference on Research in Adaptive and Convergent Systems, RACS 2015, pp. 406–411. ACM (2015)
11. Rashid, R., Tevanian, A., Young, M., Golub, D., Baron, R., Black, D., Bolosky, W., Chew, J.: Machine-independent virtual memory management for paged uniprocessor and multiprocessor architectures. In: Proceedings of the Second International Conference on Architectual Support

for Programming Languages and Operating Systems, ASPLOS II, pp. 31–39. IEEE Computer Society Press, Los Alamitos, CA, USA (1987). doi:10.1145/36206.36181

12. Saxena, M., Swift, M.M., Zhang, Y.: Flashtier: A lightweight, consistent and durable storage cache. In: Proceedings of the 7th ACM European Conference on Computer Systems, EuroSys'12, pp. 267–280. ACM, New York, NY, USA (2012). doi:10.1145/2168836.2168863

13. Shapiro, J.S., Adams, J.: Design evolution of the eros single-level store. In: Proceedings of the General Track of the Annual Conference on USENIX Annual Technical Conference, ATEC '02, pp. 59–72. USENIX Association, Berkeley, CA, USA (2002). http://dl.acm.org/citation.cfm?id=647057.713855

14. Shvachko, K., Kuang, H., Radia, S., Chansler, R.: The hadoop distributed file system. In: Proceedings of the 2010 IEEE 26th Symposium on Mass Storage Systems and Technologies (MSST), MSST '10, pp. 1–10. IEEE Computer Society, Washington, DC, USA (2010). doi:10.1109/MSST.2010.5496972

15. Ueda, K., Nomura, J., Christie, M.: Request-based device-mapper multipath and dynamic load balancing. In: Proceedings of the Linux Symposium, vol. 2, pp. 235–243 (2007)

How to Build a High Quality Mobile Applications Based on Improved Process

Haeng-Kon Kim and Roger Y. Lee

Abstract Mobile application development process can be tremendously rewarding for aspiring software creators. But without proper planning and resource allocation, it can also be extremely difficult. In fact, development teams often overcompensate in the early stages by devoting a great deal of time to ideation and design while forgetting to put an equally important amount of effort into back-end operations, including key stages of the process such as maintenance and support. By in the context of mobile applications development project, the applications quality can be explained as following. A project manager must estimate quality while a project is in procedure. The undisclosed matrixes collected individually by a software engineer are integrated in order to determine the project level. Although it is possible to measure various quality elements with merely the collected data, it is also required to integrate the data in order to determine the mobile applications project level. The key task during the project is to measure errors and defects, even though many quality measurements can be collected. The matrix drawn from these measurements offers the standard of the effectiveness of quality warranty and the control activity of individual and group software. In this paper, we identify defects to produce reliable mobile software and analyze the relationship among different defects among mobile applications. Another goal of this paper is to design a defect trigger based on the findings. So when we archive resembling project, we can forecast defect and prepare to solve defect by using defect trigger.

H.-K. Kim (✉)
Department of Computer Engineering, Catholic University of Daegu,
Kyung San, Daegu, Korea
e-mail: hangkon@cuth.cataegu.ac.kr

R.Y. Lee (✉)
Department of Computer Science, Central Michigan University,
Mount Pleasant, USA
e-mail: lee@cps.cmich.edu

1 Introduction

The concept of mobile software quality cannot be defined easily. Software has various quality-related characteristics. Moreover, there are various international standards for quality. In reality, however, quality management is often limited to the level of resolving defects after they occurred. Therefore, the defect density of a delivered product, that is, the number of defects or the size of defect column are used in defining the quality of a software, which has become an industry standard today. Therefore, a defect, or a software defect can be defined as the cause that makes a software work differently from its intended function according to the customers' needs and requirements.

The most important goal of software engineering is to develop products of high quality. To achieve this goal, the software engineer must apply efficient methods with cutting-edge tools throughout the software process. Moreover, the software engineer needs to confirm whether a quality product has been produced. The quality of the system, application, and the product was used to describe the requirement analysis, design, coding and test stages. An excellent software engineer evaluates the test case developed through analysis, design, source codes, and their documentation. To achieve the required quality level, the engineer must use technical measures to evaluate the quality, which relies on objective methods rather than subjective methods.

In the context of a project, quality can be explained as following. A project manager must estimate quality while a project is in procedure. The undisclosed matrixes collected individually by a software engineer are integrated in order to determine the project level. Although it is possible to measure various quality elements with merely the collected data, it is also required to integrate the data in order to determine the project level. The key task during the project is to measure errors and defects, even though many quality measurements can be collected. The matrix drawn from these measurements offers the standard of the effectiveness of quality warranty and the control activity of individual and group software.

In this paper, we identify defects to produce reliable mobile software and analyze the relationship among different defects among mobile applications. Another goal of this paper is to design a defect trigger based on the findings. So when we archive resembling project, we can forecast defect and prepare to solve defect by using defect trigger [1].

2 Related Work

2.1 Process Concepts

Process is what people do, using procedures, methods, tools, and equipment, to transform raw material (input) into a product (output) that is of value to customers.

A software organization, for example, uses its resources (people, and material) to add value to its inputs (customer needs) in order to produce outputs (software products).

Process is a sequence of steps performed for a given purpose. Software Process is a set of activities, methods, practices, and transformations that people use to develop and maintain software and the associated products. Processes exist at various levels, and serve general or specific goals. At the organization level, processes interact broadly with the environment or seek organization-wide goals; at the tactical and operational levels, processes serve specific project or functional goals; at the individual level, processes accomplish specific tasks. The process management premise is that the quality of the product (e.g. a software system) is largely governed by the quality of the process used to develop and maintain it. Process context is an organizations as systems with strategic, technical, structural, cultural, and managerial components; relation of process to other components of organizational systems; people, process, and technology as three quality leverage points; relating process and product; relating to external forces; process levels; formal and informal processes.

2.2   Process Definition

Process definition consists of adding and organizing information to a process model to ensure it can be enacted. A process is defined when it has documentation detailing what is done, who does it, the materials needed to do it, and what is produced. A software process definition establishes a plan for applying tools, methods, and people to the task of software development. Process definition activities are product planning, process familiarization, customer identification, interviewing, analysis, model construction, verification and validation. Components of software definition document will consist of information about work product, activity, and agent viewpoints. That is, the document identifies work products to be produced, activities, and the agents involved in producing the work products. Related terms and concepts are as followings; process design, process management principles, life-cycle-models, descriptive modeling, prescriptive modeling, organizational process asset, perspective viewpoint, process asset, process model, process guide. Software processes have been categorized and structured in different ways. Two major process breakdowns are described below [2].

2.3   Mobile Applications Defect

A mobile applications defect is an important diagnostic type representing the process and product. Since a defect is closely related to the quality of software, defect data is more important than a man-month estimation in several ways.

Furthermore, defect data is indispensable for project management. Large-scale projects may include several thousands of defects. These defects may be found in each stage of the project by many people. During the process, it often happens that the person that identifies and reports a defect is not the same person who corrects the defects. Usually, in managing a project, efforts will be made to remove most or all of the defects identified before the final delivery of the software. However, this will make it impossible to report and resolve a defect in a systematic manner. The use of an unofficial mechanism in removing defects may make people forget about the defect that was discovered. As a result, the defect may remain unresolved, or additional man-month may be required later. Therefore it is a basic requirement to record and trace the defects until they are solved. For this procedure, we will need information such as the symptom of the defect, position of the suspected defect, the discoverer, the remover and so on. Since a defect can be found in the work product of a project, it may have a negative effect on achieving the goal of project. The information on defects includes the number of defects that are identified through various defect detection activities. Therefore, all defects that are discovered in the requirement examination, design examination, code examination, unit test, and other stage are recorded. The distribution data of the number of defects found in different stages of the project are used in creating the Process Capability Baseline. Several explanations are also recorded in the PDB(Process Data Base) input item, which includes explanation on estimation and risk management.

2.4   Impact of Defect

By recording the identified defects, it is possible to focus on analyzing the number of defects, the defect ratio in the initial state and other issues of the project. Such defect traceability is one of the best implementation guidelines for project management.

The analysis is made based on the types required for making use of the defect data. In this case, additional measures must be taken besides merely recording defects. For example, in order to understand the software process level, it is necessary to distinguish defects detected during test activity and those discovered after shipping. This kind of analysis becomes possible if the stages in which the defects were discovered are differentiated in order to make the base line to record the location and percentage of the defects and to compare predicted defects and actually identified defects.

One of the goals of such organizing is to improve quality and productivity through the continuous enhancement in process. Such approaches to improve quality and productivity include studies on the efficiency of defect elimination in various stages and on the possibility of additional improvement.

The DRE (Defect Removal Efficiency) of the defect detection stage is the ratio of the number of defects discovered at a stage against the total number of defects that appears when the stage is being processed. The more effective the DRE, the lower the possibility of undetected defects. This shows that increasing the DRE(Defect Removal Efficiency) is a method to improve productivity. It is necessary to identify the location of the detected defect and the time the defect is inserted. In other words, it is crucial to understand the information of the stage in which the defect was inserted regarding each recorded defect [3].

2.5 Mobile Development Process Spiral [4]

The complexity of the human side in human computer interaction makes it almost impossible to create a mobile app with few usability problems at the first trial. Combining the spiral model development process and the steps to assess the development is a novel approach which may reduce the number of usability problems in mobile apps. Each iteration of the spiral model has four parts namely requirements definition and analysis, design, implementation and testing and planning the next iteration. With the progress in this spiral, more usability details will appear. Usability assessments, in the spiral model, are iterative and use a reference in time to compare the results and amend and improve the metrics. In the following sections, we present the spiral mobile development process and the main activities in each phase incorporating the usability metric development and assessment.

The steps of developing the metrics and assessments are integrated as follows:
First iteration:

1. ***Determine requirements***: The phase starts by collecting system requirements and identifying all users, tasks and contexts in which the application will be used. Then, defining and prioritising the attributes of usability and identifying a metric to measure each attribute; specifying an ideal and acceptable value for each metric;
2. ***Design***: The first iteration of the design will be at the level of developing and sketching a low fidelity prototype of the app interface;
3. ***Test***: Upon development of the prototype, the developers will use different usability techniques to measure the actual value of each attribute and calculate the rating of each attribute;
4. ***Plan*** next iteration

Second Iteration:

1. ***Determine requirements***: in this phase, the developing team will have a better idea of what the usability requirements are for the system although the design requirements will not be complete. In that respect, step 1 will be repeated and the developers will collect more data and app requirements and explore whether there are more potential users, tasks and contexts in which the application will be used. Then, the attributes of usability are redefined and prioritised.

As a result, developers alter the metrics to accommodate the added requirements and specify an ideal and acceptable value for each metric;

2. *Design*: at that level, developers start to develop a high fidelity prototype focusing on the interface.
3. *Test*: Upon development of the prototype, using the usability techniques the actual value of each attribute will be measured and the rating of each attribute will be calculated and compared with the previous iteration results.
4. *Plan* next iteration

Third Ieration:

1. *Determine requirements*: In this iteration, most of the design requirements are becoming more clear and by using the earlier iteration results, developers can identify and prioritise the attributes of usability, define a metric to measure each attribute and specify an ideal and acceptable value for each metric;
2. *Design*: here the whole system is developed and the alpha version is realized.
3. *Test*: Upon development of the alpha version, the developing team use usability techniques and measure the actual value of each attribute, calculate the rating of each attribute and compare rating with the previous iteration.
4. *Plan* next iteration

Fourth Iteration:

1. *Determine requirements*: In this phase, the results of the previous iteration are used to identify and prioritise the attributes of usability and define a metric to measure each attribute and specify an ideal and acceptable value for each metric;
2. *Design*: with the results, the Beta version is developed and released to be evaluated
3. *Test*: Again using the usability techniques we measure the actual value of each attribute, calculate the rating of each attribute and compare the rating with the previous iteration.
4. *Plan* next iteration

Fifth Iteration:

1. *Determine requirements*: the results of the previous iteration are used to identify and prioritise the attributes of usability and define a metric to measure each attribute and specify an ideal and acceptable value for each metric;
2. *Design*: final product is developed
3. *Test*: a usability assessment is done using the measure of the actual value of each attribute, calculate the rating of each attribute and compare the rating with the previous phase. An alteration to the final product is done based on the results and released to the product
4. *Plan* a final report of the results is issued

Our novel spiral mobile development process makes an expensive use of usability techniques. Therefore, the following section lists the most commonly used and recommends a few of them. The Usability literature offers numerous techniques to be used for different project characteristics and for different usability purposes. However, we have chosen the ones that can be applied with moderate usability training and the choice will be based on the application and the context of use. Table 1 summarizes these techniques and structures them according to the Usable Spiral Mobile Development Process iterations and phases and where they can be applied (Fig. 1) [4].

2.6 Mobile Applications Defect Tree Analysis

The Mobile Applications Defect Tree analysis builds a graphical model of the sequential and concurrent combinations of events that can lead to a hazardous event or system state. Using a well-developed defect tree, it is possible to observe the consequences of a sequence of interrelated failures that occur in different system components. Real-time logic (RTL) builds a system model by specifying events and

Table 1 Mobile applications questionnaire structure in requirements analysis stage

Life cycle	Detailed category	Question purpose
Requirement analysis	Completeness	Questions asking whether the algorithm is defined to satisfy validity of requirement
	Correctness	Questions examining whether requirement is embodied by function
	Quality attribute	Questions about items related to quality attribute to confirm whether requirement is satisfied
	Traceability	Questions to confirm consistency of requirement
	Etc.	Questions for other items that do not belong in the detailed category

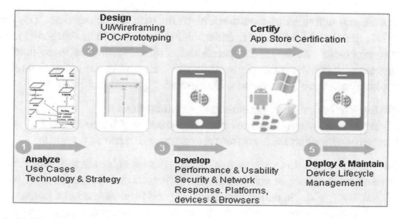

Fig. 1 Mobile development process spiral

corresponding actions. The event-action model can be analyzed using logic operations to test safety assertions about system components and their timing. Petri net models can be used to determine the defects that are most hazardous. If a risk is identified and analyzed, the requirement for safety can be specified on the software. In other words, the specification can include a list of undesired events and anticipated system responses. This is related with the software's role of managing undesired events. Although software reliability and its safety are closely inter-related, it is important to understand the subtle difference between them. For determining software reliability, statistical analysis is used to calculate the probability of a software defect occurrence. However, the occurrence of a defect does not necessarily lead to a danger or disaster. Mobile Applications safety may be measured by the extent that a failure actually results in a disaster.

In other words, a failure is not evaluated in a vacant or isolated state but within the entire computer-base system [5, 6].

3 How to Apply Improved Process for Quality Mobile Applications

This chapter will develop a questionnaire to identify defects that occur during actual projects of companies. Therefore, the structure and contents of the questionnaire to detect the defect and analyze the result are presented in this chapter. Furthermore a defect trigger based on the analyzed result is provided. As for the domain of the project, digital systems were selected and 21 projects of four companies were chosen as target companies [7].

3.1 Collecting Defect Data

The operational activities of measurement begin with collecting data. The procedures that are defined for collecting and retaining data need to be integrated into the software processes and be made operational. Collecting data is more than mere measurements. It consists of implementing plans, sustaining the research work, and the resulting measurement activities. Documenting procedures involves:

Identifying the responsible persons and organizations
Specifying where, when, and how measurements will be conducted
Defining procedures to be used for recording and reporting results

The requirements help to ensure that the assessment output is self-consistent and that it provides evidence to substantiate the ratings. The assessment shall be conducted following a documented process that meets the assessment purpose. The assessment process shall contain at minimum activities such as planning, data collection, data validation, process rating, and reporting. The training and assessment of the ISO Guide 62 and TR 15504, part 6 supported the reliability of the assessment. Furthermore, the people who rated the maturity questionnaire have been screened in order to confirm the reliability of the assessment.

The certified assessors who conducted assessments of the concerned company with SPICE (ISO TR 15504, ver.3.3, 1998) also rated the SEI maturity questionnaires.

The project portfolio assessed is in the following table.

3.2 Questionnaire to Detect Defects

A defect detection questionnaire was made to identify defects in the ongoing project of a company. Another purpose is to discover defects and analyze the causes between each defect. Questions that must be examined in the questionnaire were divided as below by classifying defects that exist in each life cycle into a detailed category. The structure of the questionnaire for each life cycle is as follows (Table 2).

To detect defects, five detailed categories were made for the questionnaire at the requirement analysis stage. Detailed categories for defect detection at the requirement analysis stage reflect customers' requirements and confirm how the requirements are satisfied. The requirements are classified in order to confirm consistency. Therefore, the detailed category consists of completeness, correctness, quality attribute, and traceability. For other items, an investigation was made in order to

Table 2 Mobile applications questionnaire structure in design stage

Life cycle	Detailed category	Question purpose
Design	Structure	Questions asking whether the structure in the design phase is clear, and whether integration and examination are easy
	Data	Questions asking whether data is initialized to precisely define relation between modules (components)
	Correctness	Questions asking whether designing is carried out to satisfy requirement
	Standard and traceability	Questions asking whether a standard process is used and whether the requirement stage is traceable during design
	Logic	Questions asking whether there is no logical error
	Interface	Questions asking whether interface is precisely defined
	Clearness	Questions to ask whether consistency exists in product present in the design phase and is clearly expressed
	Robustness	Questions asking whether processing in an exceptional situation is considered

Table 3 Mobile applications questionnaire structure in coding stage

Life cycle	Detailed category	Question purpose
Coding	Method	Question asking whether the function is completely realized in implementation
	Static relation	Question asking whether error exists in the internal contents of code
	Dynamic relation	Question asking whether error is detected within connection of outside device and the relevant file is managed

Table 4 Mobile Applications Questionnaire Structure in Test Stage

Life cycle	Details Category	Question purpose
Test	Test plan	Questions asking whether test environment and objectives are planned by schedule
	Correctness and completeness	Questions asking whether guidelines and standard are clearly defined for a precise and perfect test implementation
	Standard and traceability	Questions asking whether test is being implemented based on requirements
	Regression test	Questions asking whether change and version of code are differentiated and are modified
	Resources and schedule	Questions asking whether responsibilities and roles are assigned based on a scheduled resource allocation

confirm whether the specification was based on the planned form and schedule that support the four-detailed category. Each detailed category is shown in Table 3.

The structure of the Defect detection questionnaire in the design stage is shown in Table 4. The purpose of developing a detailed category is to confirm that customer's requirement in the design stage is well reflected, which is a prerequisite for the requirement to be correctly reflected in the implementation stage. Therefore the detailed categories were divided according to how clearly the module specification in the design stage was defined both internally and externally. The category is composed of structure, data, correctness, standard, traceability, logic, interface, clearness and robustness.

The questionnaire structure in the coding stage is presented in Table 5. Classification of a detailed category confirms how the requirement is implemented reliably as a function in the implementation stage after it has gone through design stage.

When an error was identified by analyzing the relations between code interior and exterior, the detailed category was classified based on standards confirming how this error was detected and resolved.

The questionnaire structure in the test stage is presented in Table 6. The purpose of the detailed category of the questionnaire is to confirm that the test is implemented on schedule, and under the planned environment and standard. Another

Table 5 Mobile applications detected defects in requirements analysis stage

Category	Question contents
Completeness	1. Is the written requirement described correctly in detail?
	2. Was the basic algorithm defined to satisfy functional requirements?
	3. Does software requirement specification include all the requirements of the customer and system?
	4. Was mutual confirmation of conformability with other requirements made?
Correctness	1. Is a consistent delivery of a specified error message possible?

Table 6 Mobile applications defects detected in design stage

Category	Question contents
Data	1. Did you explain the content that was omitted in description of system data?
	2. Was all the data defined properly and was initialized?
Correctness	1. Are reliability and performance requirements specified?
	2. Did undefined or unnecessary data structure exist?
	3. Did you consider all constraint items?
	4. Can you analyze to decide required productivity, response time, and correctness? Can you implement this in design?
	5. Can you verify in design stage?
Logic	1. Is there missing or incomplete logic?
Interface	1. Are all interfaces defined clearly?

purpose is to confirm whether the test result satisfies customers' requirements and whether correct responsibilities and roles are assigned based on a planned resource allocation.

The actual questionnaire includes 17 requirements, 28 designs, 11 items for coding and 26 tests. The surveyed person is required to answer the questions asking whether the function works correctly with "Yes" or "No". Even when the response is "Yes", a problem may occur later even if the function or standard asked exist.

Therefore, such cases were classified as a defect, and the ratio of the defect of the relevant question against total defects for each question item was calculated.

3.3 Mobile Applications Data Analysis of Questionnaire to Detect Defects

Among the question items in the defect detection questionnaire, there were items to which developers commonly replied that there was a defect. These items were classified separately. When these question items are classified according to life cycle and category, they are as follows. The analysis of the questionnaire in which defects existed in the requirement analysis stage is as follows. Each company was found to conduct design and implementation stages by specifying extracted requirements. However, it has been discovered that even if the requirements change and are functionalized, the task to check whether the intended functions are satisfied seemed insufficient. According to the question results, most companies were found to reuse the extracted requirements without modification. Therefore, methods to extract the customer's requirement and to confirm whether there are additional changes in the requirement and to confirm whether the requirements are met are required.

Defects detected in the requirement stage are more important than defects detected in other development stages.

The defect ratio was 35% against the total, of which 25% was for completeness, and 10% for correctness. Although defects were also identified in the design, implementation and examination stages, it has been found out that that defects in the requirement stage were responsible for the defects in the following stages. Therefore, predicting defects and finding a solution in the initial stage of a project would have a positive effect in terms of schedule, cost, and quality of the project.

The contents of the defect detection questionnaire in the design stage are as follows. The design stage is a stage to systemize the extracted requirements so that they are functionalized in the system.

From analyzing the questionnaire answers, it is possible to say that a defect existed in the base structure for using the data in the design stage. Furthermore, in order to predict the mutual relationship among modules, it is necessary to clearly define the interface. The defect ratio in the design stage was 25% against the total defects according to the analysis. Among this 25%, 10% was for correctness, 5% for data, 5% for logic, and 5% for interface. The questionnaire for defects detected in the design stage is shown in Table 7.

Defects in the coding stage were mainly functional errors that happen in a code. However, as the products near delivery to customers as each stage are processed after the coding stage, the companies analyzed the stages after coding in detail and prepared alternatives. However, defects that resulted from the defects in the requirement analysis and the design stage took up a large portion among defects in the coding stage (Table 8).

The defect ratio in the coding stage was 15% out of total defects, among which 5% was for method, 5% for static relation, and 5% for dynamic relation [8, 9].

Table 7 Mobile applications defects items detected in coding stage

Category	Question contents
Method	1. Is function associated with outside spec processed correctly?
	2. Is the code re-useable by calling external reusability component or library function?
	3. Can you make one procedure by summarizing recursive code block?
Static relation	1. Is code documented in comment form for easy management?
Dynamic relation	1. Is the existence of a file confirmed before accessing the file?

Table 8 Table of defect removal efficiency in mobile applications

	Number (%) of defect found at relevant S/W development step (E)	Number (%) of defect found at next S/W development step (D)
Requirement	10	3
Design	15	3
Coding	5	1

Table 9 Size based software metrics in mobile applications

	Last project	The present project
SLOC	40,000	120,620
Project cost	400,282,000	1,500,000,000
Effort (Man-Month)	55.2	381.6
Defect number	400	810
Project people	11	50

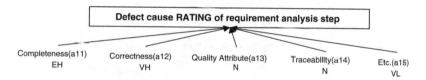

Fig. 2 Cause rating scale of defect in requirement analysis stage

The defect detection questionnaire in the coding stage is presented in Table 9.

In the test stage, defects were not applied with the tolerance limit and output standard of the defect during testing.

The resource and human resource allocation and role assignment in planning the test were not satisfactory.

According to the analysis, the defect ratio in the test stage was 25% out of total defects, among which 10% was for completeness and correctness, 5% for standard and traceability, 5% for revolution examination, and 5% for resources and schedule [10–12].

3.3.1 Mobile Applications Hierarchical Decision-Making Structure

The decision-making structure and graded items in each stage are as below (Fig. 2).

The defect cause rating scale structure is divided into four stages (requirement analysis, design, coding, test), and each stage has been further categorized in detail.

The rating scale is divided into six dimensions: Very Low (VL), Low (L), Normal (N), High (H), Very High (VH), Extra High (EH). This means that, among the six dimensions, priority increases as it nears EH, and the priority decreases when as it approaches VL.

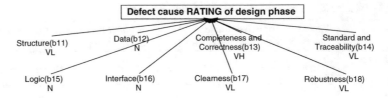

Fig. 3 Cause rating scale of defect in design stage

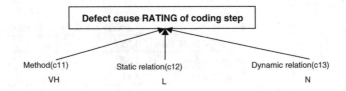

Fig. 4 Defect cause rating scale in coding stage

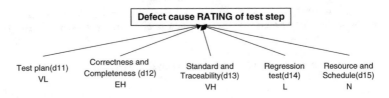

Fig. 5 Defect cause rating scale in test stage

In the case of Fig. 3, at the requirement analysis stage, a rating scale was made based on the weight of each item analyzed in 3.3. The result showed that the completeness item had the most defects.

Therefore, completeness and correctness were found out to be greater causes of defect than quality attribute and traceability.

It has been identified that the defect cause in the design stage of Fig. 4 was correctness and completeness. The defect cause in coding stage of Fig. 5 was method, and the defect cause in test stage of Fig. 6 was correctness and completeness, standard and traceability.

3.4 Mobile Applications Defect Trigger

There are many tools today that are used to define the relations among cost, schedule, and quality. Among them, those that focus on the relations among defects are quality models COCOMOII and COQUALMO, which are under extension and

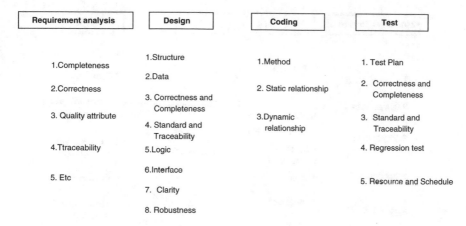

Fig. 6 Mobile applications structure of defect trigger

further development at present. Among the two, especially COQUALMO focuses on detecting and removing defects [13, 14].

A defect detection questionnaire was developed to ask questions of companies in order to analyze defects. Therefore, the defects of each development stage were detected, and the result was analyzed. Also, in order to prevent a defect, a defect trigger based on the analysis of detected defects is presented.

The defect trigger distinguishes the different causes of defects and finds the relation between defects, thereby revealing the cause of a defect. Therefore, defect data must be used to complete a defect trigger. By using a defect trigger in a similar project, it is possible to identify the relation between defects and predict possible defects, which ultimately will be a useful tool for defect prevention.

In the previous Sect. 3.4, in order to design the defect trigger, the triggers items of each development stage were presented according to a detailed category explained in Sect. 3.1. The structure of the defect trigger is as follows.

By using a defect trigger, it is possible to analyze the causes of defects in all previous stages (requirement analysis, design, coding and test stage).

According to the result of the analysis, identified the relation between defects and software defect process.

The trigger about the relation between defects is as follows (Fig. 7).

Mobile applications defect of data category in the design stage is attributable to the fact that the definition of the data content were not analyzed in detail in the requirement analysis stage. In addition, as for defects involving correctness and logic category, the error could not be passed on correctly, since the data contents were lacking. The causes of defects in the correctness category are related to completeness in the requirement analysis stage, and those in the logic category were found to be related to the correctness category.

Because the conformability between requirements was not confirmed sufficiently, the interface category could not be clearly defined. Therefore, it has been

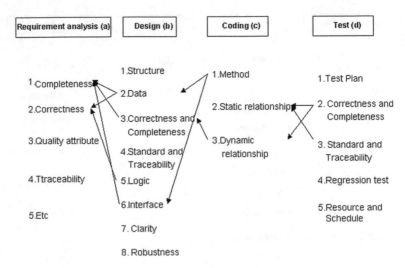

Fig. 7 Association trigger between defects

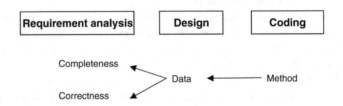

Fig. 8 Mobile applications defect Trigger-1 of details category

analyzed that the completeness category of in the requirement stage to be the causes of these defects.

The defects in the coding stage were caused by data in the design stage and interfacing problems. The reason is that there is no detailed explanation about the repetition codes and the relationship with external specification, and that the distinction of inter-relationships through interfacing is lacking. The causes of defect in the dynamic relation category were found to be the inaccuracy in the design stage due to insufficient consideration of restriction.

The defect causes of correctness and completeness category in the test stage were due to insufficient consideration of documentation and constraint.

Therefore, the defect causes of the correctness and completeness category at the coding stage were attributable to faults in the static and dynamic relation category. The defect causes of the standard and traceability category resulted from faults in the static relation category at the test stage, due to lacking documentation.

In conclusion, by removing defects in the completeness and correctness category at the requirement stage, the defects of data, correctness, logic and interface can be prevented at the design stage. In this regard, when expressing a causal relationship in the trigger in a picture, it is as follows (Figs. 8, 9, 10 and 11) [15, 13].

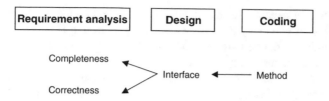

Fig. 9 Mobile applications defect Trigger-2 of details category

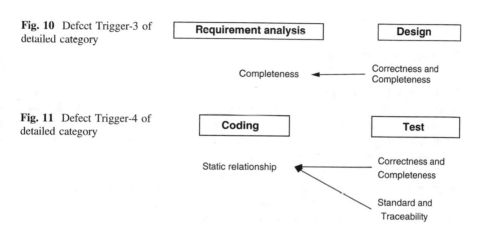

Fig. 10 Defect Trigger-3 of detailed category

Fig. 11 Defect Trigger-4 of detailed category

4 Effectiveness Analysis of Mobile Applications Defect Trigger

We analyzed aspect that defect removal is achieved efficiently to analyze effectiveness of defect trigger.

Also, we present effectiveness of introduction through if productivity improves because defect of whole project is reduced through defect trigger.

4.1 Defect Removal Efficiency Analysis on Mobile Applications

Purpose of Defect Trigger design improves quality of product and heighten productivity.

Therefore, when we applied Defect Trigger in actuality project, we wish to apply defect exclusion efficiency (Defect Removal Efficiency). To measure ability of defect control activity.

After apply Defect Trigger, defect exclusion efficiency analysis investigated defect number found at relevant S/W development step and defect number found at next time step in terms of request analysis, design and coding stage. Production of defect exclusion efficiency is as following. DRE = E(E + D).

E = Number of defect found at relevant S/W development step (e.g.: Number of defect found at request analysis step).

D = Number of defect found at next S/W development step (e.g.: Defect number that defect found at design step is responsible for defect of request analysis step).

Ideal value of DRE is 1, and this displays that any defect does not happen to S/W [16].

5   Conclusions

In the context of this project, quality can be explained as following. A project manager must estimate quality while a project is in procedure. The undisclosed matrixes collected individually by a software engineer are integrated in order to determine the project level. Although it is possible to measure various quality elements with merely the collected data, it is also required to integrate the data in order to determine the project level. The key task during the project is to measure errors and defects, even though many quality measurements can be collected. The matrix drawn from these measurements offers the standard of the effectiveness of quality warranty and the control activity of individual and group software. In this paper, we identify defects to produce reliable mobile software and analyze the relationship among different defects among mobile applications. Another goal of this paper is to design a defect trigger based on the findings. So when we archive resembling project, we can forecast defect and prepare to solve defect by using defect trigger. In implementing a similar project, using such triggers will help produce a more reliable result in terms of cost, quality, and schedule of the entire project by predicting the stages where defects occur and the contents of the defects. In order to make this possible, further study is required on how this can be used to enhance the COQUALMO defect reduction estimates by using the trigger information. The direction of further study should be to develop a defect trigger framework by providing detailed items of the defects, and to develop a system that is capable of predicting defects by inputting major items on the mobile applications and clouding computing.

Acknowledgements This research was supported by Catholic University of Daegu.

This research was also supported by the International Research & Development Pro-gram of the National Research Foundation of Korea (NRF) funded by the Ministry of Science, ICT & Future Planning (Grant number: K 2014075112).

References

1. Aberdeen Group, Mobile Retail is a Reality: The Increasing Mobility of Consumers has Retailers Engaged, Aberdeen's Insights (2010). http://www.mobilemarketer.com/cms/lib/9806.pdf.
2. ABI Research, Technology Market Intelligence, One Billion Mobile Broadband Subscriptions in (2011) http://www.abiresearch.com/press/3607-ne+Billion+Mobile+Broadband+Subscriptions+in+2011:+a+Rosy+Picture+Ahead+for+Mobile+Network+Operators accessed in, (2011) November.
3. Nebe, K., Paelke, V.: key requirements for integrating usability engineering and software engineering,, HCII 2011 Orlando, Florida, USA, vol. 6761, pp. 114–120. Springer-Verlag (2011).
4. Abrahamsson, P., Hanhineva, A., Hulkko, H., Ihme, T., Jäälinoja, J., Korkala, M., Koskela, J., Kyllönen, P., Salo, O.: Mobile-D: an agile approach for mobile application development," in ACM, 19th Annual ACM Conference on Object-Oriented Programming, Systems, Languages, and Applications (OOPSLA) Vancouver, Canada, October (2004).
5. Morris, B., Bortenschlager, M., Luo, C., Sommerville MLansdell, J.: An Introduction to Bada: A Developer's Guide" WILEY, West Sussex: UK (2010).
6. Rahimian, V., Ramsin, R.: Designing an agile methodology for mobile software development: a hybrid method engineering approach, in IEEE, 2nd International Conference on Research Challenges in Information Science (RCIS)Annual Conf. Marrakech, Morocco, pp. 337–342, June (2008).
7. Software Process Improvement Forum, KASPA SPI-7, (2002) December.
8. Chillarrege, R., Prasad, K. R.: Test and Development Process Retrospective? a Case study using ODC Triggers", IEEE computer Society, (2002), 4.
9. N. Fenton and N. Ohlsson "Quantitative analysis of faults and failures in a complex software system, IEEE Trans. Software Eng., vol. 26, pp. 797–814, (2000).
10. SPICE Assessment in korea, The korea SPICE, (2002) May 13.
11 Shin, K. A.: Research about software reliability development model that defect importance is considered", Journal of the korea computer Industry education society, vol 03, no. 07, pp. 0837–0844, (2000).
12. Dunn, R. H.: Software defect removal, McGraw-hill, (1984).
13. Mladen, V. A.: Software Reliability Engineering, Tutorial Notes? Topics in Reliability & Maintainability & Statistics, 2000 Annual Reliability and Maintainability Symposium, Los Angeles, CA, (2000) January 24-27.
14. Padberg,: A Fast Algorithm to Component Maximum Likelihood Estimates for the Hypergeometric Software Reliability Model", Asia-Pacific Conference on Quality Software APAQS, vol. 2, pp. 40–49, (2001).
15. McCall, P. K., Richards and G. F. Walters, Factors in software quality, vol. 1, 2 and 3. Springfield VA., NTIS, AD/A-049-014/015/055, (1997).
16. Runeson, W.: Defect Content Estimations from Review Data, Proceedings International Conference on Software Engineering ICSE 400–409, (1998).
17. Nosseir, A., Flood, D., Harrison R., Ibrahim O.: Mobile Development Process Spiral, Research paper, British University in Egyp Dundalk Institute of Technology.
18. Zhang, D., Adipat, B.: Challenges, methodologies, and issues in the usability testing of mobile applications." International Journal of Human-Computer Interaction 18th ed, vol. 3, pp. 293–308, (2005).

A New Hybrid Discrete Firefly Algorithm for Solving the Traveling Salesman Problem

Abdulqader M. Mohsen and Wedad Al-Sorori

Abstract Firefly algorithm is a new meta-heuristic inspired by a natural phenomenon of fireflies' flashing light. Firefly algorithm has been successfully applied to solve several optimization problems. However, It still suffers from some drawbacks such as easily getting stuck at local optima and slow speed of convergence. This paper proposes a new hybrid variant of discrete firefly algorithm, called HDFA, to solve traveling salesman problem (TSP). In the proposed improvement, the balance between intensification and diversification is achieved by utilizing the local search procedures, 2-opt and 3-opt, to improve searching performance and speed up the convergence. In addition, the genetic algorithm operators, crossover and mutation, are added to allow performance of both local and global search respectively. The validity of HDFA is verified by comparative experiments using eighteen TSP benchmark instances from TSBLIB and compared to some well-known algorithms. Results in the conducted experiments show that HDFA has significantly better performance than the performance of compared algorithms for all instances in terms of solution quality.

1 Introduction

The traveling salesman problem (TSP) is an optimization problem that was proposed in the early 1930 s by Karl Menger, and later was promoted by Hassler Whitney and Merrill Flood. Given a set of cities and the cost of traveling between each possible pairs, TSP aims to find the best possible route to visit all cities and return to the starting point with minimum traveling cost. The total number of possible routes covering all n cities can be given as a set of feasible solutions equal to $(n - 1)!/2$ [1].

A.M. Mohsen · W. Al-Sorori (✉)
University of Science and Technology, Sana'a, Sana'a, Yemen
e-mail: w.alsrori@ust.edu

A.M. Mohsen
e-mail: a.alabadi@ust.edu

© Springer International Publishing AG 2017
R. Lee (ed.), *Applied Computing and Information Technology*,
Studies in Computational Intelligence 695,
DOI 10.1007/978-3-319-51472-7_12

In 1972, Karp proved that TSP is an NP-hard problem (NP stands for nondeterministic polynomial time). TSP was used as a benchmark for many optimization problems to find out the optimal solutions with minimum cost [2]. All known exact methods for TSP such as dynamic programming are time consuming and require an exponentially increasing amount of time; which takes an order of $O(n^2 2^n)$ [3]. This motivated researchers to develop various meta-heuristic algorithms [4].

Firefly Algorithm (FA) is a swarm based meta-heuristic algorithm introduced by yang in 2008. FA was inspired by tropical fireflies' communication system in which fireflies using flashing light to attract mating partners. Like other swarm based metaheuristic algorithms, FA was initially developed to solve continuous optimization problems then transferred to solve different optimization problems from different fields. FA success attracted researchers' attention to make more improvements on the basic FA structure which resulted in introducing several variants as shown in surveys [5, 6]. Most of these variants went to hybridize FA with other techniques. Hybridization of FA with other algorithms may increase the chances of finding good quality solutions compared with the original FA. For solving the continuous optimization problems, FA was hybridized with different methods such as Rampriya et al. [7], Farahani et al. [8], Abdullah et al. [9], Rajini and David [10], and Yang and Deb [11]. For combinatorial problems, FA also looks to be a promising algorithm for solving problems from this category such as scheduling and TSP. Thus, different methods were hybridized with FA to improve its performance.

To solve TSP, FA was adapted and hybridized with different methods. For example, an evolutionary discrete firefly algorithm (EDFA) was introduced, by Jati and Suyanto [12], in which mutation and selection operations were incorporated with FA. EDFA suffered from the drawback of trapping in local optima in some of its instances. Then Saraei et al. gained good results somewhat by integrating greedy approach with FA but it took long time to obtain the optimality [13]. Zhou introduced a multi-population FA with k-opt algorithm, called MDFA, [14, 15]. He proved that MDFA outperformed the previous FA variant named EDFA in terms of convergence rate and solution accuracy. In spite of using small-scale data sets to prove its superiority, MDFA still suffers from trapping in local optima. Motivated by this, we proposed a new discrete FA variant to solve TSP in such a way that FA was hybridized with two genetic operators i.e. mutation and crossover and two local search procedures i.e. 2-opt and 3-opt. The crossover and mutation operators were incorporated to strike a good trade-off between diversification and intensification and therefore to overcome the FA limitations of stagnation and getting stuck in the local optima. Meanwhile, 2-opt and 3-opt were added to speed up the convergence.

In addition to this section, this paper was organized as follows. Section 2 provides an overview for the paper key concepts such as the basic FA, Discrete FA variant and crossover and mutation operators. Section 3 presents the proposed algorithm, HDFA, in details. Section 4 shows and discusses the experimental results. Section 5 draws the conclusions and summarizes some possible future works.

2   Preliminaries

In the following subsections, we briefly describe paper's key concepts i.e. the firefly algorithm, FA discrete variant, crossover and mutation operators.

2.1   Firefly Algorithm

A new swarm based meta-heuristic algorithm called firefly algorithm was proposed by Xin-She Yang [16, 17]. FA was inspired by the fireflies' communication system using the flashing light. It exploits the inverse-square law for light intensity in which the intensity I is inversely proportional to the square of the distance from the light source taking into account the intensity reduction due to the medium absorption. Basic FA flow can be summarized as the following steps:

Step1. (Population Initialization): Generate initial population of fireflies randomly considering all values from the possible range of values to ensure a wide variety of solutions.

Step 2. (Evaluating Fireflies): Calculate the light intensity I for each firefly in the population and evaluate that firefly according to the fitness function. Light intensity $I(x_i)$ of firefly i at position x_i is proportional to the value of its fitness function $I(x_i) \propto f(x_i)$, light intensity varies according to Eq. 1:

$$I = I_0 \exp - \gamma r^2, \tag{1}$$

where I_0 indicates the source of light intensity, whilst γ represents the fixed light absorption coefficient.

Step 3. (Population Updating): Generate a new brighter firefly for each firefly in the population according to the attractiveness between them. The attractiveness is proportional to the light intensity and is defined as shown in Eq. 2:

$$\beta = \beta_0 \exp - \gamma r^2, \tag{2}$$

where β_0 is the attractiveness at distance $r = 0$. The movement from firefly i toward firefly j can be computed as shown in Eq. 3:

$$x_i^{t+1} = x_i^t + \beta_0 \exp - \gamma r_{ij}^2 (x_j^t - x_i^t) + \alpha \epsilon_i^t, \tag{3}$$

where α is a randomization parameter, ϵ_i^t is a vector of random numbers at iteration t and γ is the light absorption coefficient. The Cartesian Distance between the two fireflies i and j at positions x_i and x_j is denoted by r_{ij}. This distance is calculated according to Eq. 4:

$$d_{ij} = \| x_i - x_j \| = \sqrt{\sum_{k=1}^{k=N} (x_{ik} - x_{kj})^2}, \tag{4}$$

where α_0 is an initial randomness factor that is reduced gradually to ensure the proper algorithm convergence.

Step 4. (Find the Current Best Firefly): Determine the current best firefly among all fireflies in the population at iteration t.

Step 5. (Termination Criterion Checking): Repeat steps 2, 3 and 4 until the required number of iterations or the optimal solution is reached.

Return the global best firefly x^* among all fireflies in the population.

2.2 Discrete Firefly Algorithm for TSP

FA was adapted to solve TSP where each firefly is represented as a permutation of all cities in the tour. To represent the fireflie's movement, Zhou et al. adapted FA with the k-opt move [15] in which the distance between two fireflies i and j was calculated using the Hamming Distance (HD).

The same movement idea was also used with improved bat algorithm (IBA) [18]. As such, each firefly i will move from one position $x_i - 1$ to a new position x_i at iteration t according to Eqs. 5 or 6:

$$x_i^t = 2 - opt(x_i^t - 1), \tag{5}$$

$$x_i^t = 3 - opt(x_i^t - 1), \tag{6}$$

This movement is controlled by the light intensity value of the fireflies. The light intensity I is computed using Eq. 7:

$$I_i = random(1, r_{ij}), \quad r_{ij} = HD(x_i^t, x^*) \tag{7}$$

where I_i is the light intensity of the firefly i that will be randomly selected from 1 to r_{ij}. r_{ij} is the Hamming Distance between firefly i and the best firefly in the population j. According to the light intensity value, each firefly will perform either short or long move using 2-opt and 3-opt local search procedures respectively as shown in Eqs. 5 and 6.

2.3 Crossover Operator

Crossover operator is used by genetic algorithm to recombine two solutions to get a better solution. Thus, it is an effective operation to exploit the solutions of search space for local search. In the literature, different types of crossover operators were introduced in [19]. One of which is the partially matched crossover (PMX) that is widely used for problems with permutation representation such as TSP. In PMX, two solutions from the population are selected as parents and then two crossover points are chosen randomly from the two selected solutions. The part of solutions between the two crossover points gives a matching selection, called swab, which affects cross through position-by-position exchange operations. For example, given the following two parents:

P1: 1 2 3 | 4 5 6 7 | 8 9

P2: 4 5 2 | 1 8 7 6 | 9 3

Then the two generated offspring when applying PMX crossover are:

O1: 4 2 3 | 1 8 7 6| 5 9

O2: 1 8 2 |4 5 6 7 | 9 3

The step by step process of PMX crossover is illuminated in Algorithm 1.

Algorithm 1: PMX crossover algorithm for TSP problem

---

1 Select the start crossover point cp1 randomly in the range [0 , *num_of_cities* − 1];
2 cp2 =(cp1 + *current_lightintensity*)%(*number_of_cities* − 1);
3 parent1=x*;
4 parent2=x_i;
5 Copy the swab, i.e., the region between cp1 and cp2, from parent1 into the first offspring;
6 **for** *each index i in the swab* **do**
7   **if** *parent2[i] ≠ offspring[i]* **then**
8   add parent2[i] to the set Elements;

9 **for** *each item e in Elements* **do**
10   Find index i of e in parent2;
11   Find the item e' in parent1 in index i;
12   Find e' in parent2 and its index k;
13   **if** *index k in parent2 is in the swab* **then**
14   e=e';
15   **go to** 10;
16   **else**
17   offspring[j]= e;

18 Fill all the remaining positions of the first offspring from parent2.
19 Analogously, **go to** 5 to create second child using parent1=x_i and parent2=x*;

2.4 Mutation Operator

Mutation operator is also used by genetic algorithm to maintain the diversity of the population and prevent the solutions in the next generation from converging to local minima. In mutation, each solution in the population may modified to produce new solution with a given probability of mutating. Adding mutation operator to FA may cause the fireflies positions to be different from their parent positions and thus explore new areas in the search space. The interchanging mutation (IM) was commonly used for problems with permutation representation such as TSP. To explore new possible solutions, IM is applied for TSP by selecting two positions of two cities randomly, in the tour, for exchange to produce new different solution. The main steps of IM are shown in Algorithm 2.

Algorithm 2: Algorithmic steps of interchanging mutation operator for TSP problem

---

1 **begin**
2   **Step1**: For solution X_i, select two cities, m_1 and m_2, in the solution at random;
3   **Step2**: Swap $X_i(m_1)$ and $X_i(m_2)$;
4   **Step3**: Return new mutated solution X'_i;
5 **end**

---

3   Hybrid Discrete Firefly Algorithm for TSP (HDFA)

In this section, a new hybrid discrete FA, called HDFA, is introduced. The main purpose of HDFA is to overcome the shortcomings of firefly algorithm mentioned above by adding components from other algorithms. Two local search procedures i.e., 2-opt and 3-opt are introduced to speed up the convergence. In addition, crossover and mutation operators of genetic algorithm are combined to achieve a balance between diversification and intensification. Thus, HDFA can maintain the diversity of the population and alleviates the occurrence of premature convergence. HDFA has 8 steps summarized as follows.

Step 1. Parameters Initialization: Initialize HDFA parameters to control its performance. These parameters are the light intensity (brightness), population size, crossover rate and the mutation rate.

Step 2. Population Initialization: Generate initial population of fireflies randomly considering all values from the possible range of values to ensure a wide variety of solutions.

Step 3. Computing the Light Intensity: Calculate the light intensity for each firefly in the population by firstly determining its HD from the current global best solution and then by choosing its light intensity I_i as a random number in the range between 1 and HD according to Eq. 8:

$$I_i = rand[1, HD(x_i^t, x^*)] \tag{8}$$

Step 4. Crossover: Apply the crossover operator between the global best firefly position x^* and the current firefly position x_i. After crossing, the fitness of current firefly position x_i is compared with the positions of the two offspring produced by crossover to select the best one as the new firefly position x'_i.

Step 5. Local Search: For each firefly i, create a new better firefly position x'_i by moving it toward the brightest one j according to their light intensity and the attractiveness between them. In this step, if the light intensity is small, i.e., $I_i \leq number\_of\_cities/2$, apply 2-opt procedure; otherwise, apply 3-opt procedure. Calculate the fitness function of the new firefly accordingly.

Step 6. *Mutation*: To increase the diversity, apply the mutation operator for each firefly in the population with a probability equals to a predefined mutation rate. In this process, if a random number in the range [0, 1] is less than the mutation rate, the algorithm alters the firefly position x_i by exchanging the content of two bits, in that firefly, selected randomly and keeps the other bits unchanged. Calculate the fitness function of the new firefly accordingly.

Step 7. *Finding the global best solution*: Determine the current global best firefly among all fireflies in the population at iteration t.

Step 8. *Termination Criterion Checking*: If optimal solution or the maximum number of iterations is reached, return the global best firefly; otherwise, repeat steps 3–7. The pseudocode and the steps of HDFA are presented in Algorithm 3.

Algorithm 3: Hybrid Discrete Firefly Algorithm(HDFA)

```
 1  begin
 2  │   Input:objective function f(x);
 3  │   Output: the best solution x*;
 4  │   t=0;
 5  │   Initialize the population p : X = x1, x2, ...., xn ;
 6  │   for i = 1 to No_of_fireflies do          /* For each firefly in p   */
 7  │   │   Calculate the objective function f(xᵢ) for firefly position xᵢ of firefly i;
 8  │   │   Light intensity Iᵢ at xᵢ is determined by f(xᵢ);
 9  │   │   Find the current best firefly x*
10  │   end
11  │   while termination criterion not reached do
12  │   │   for i=1 to No_of_fireflies do
13  │   │   │   crossover(xᵢ,x*)
14  │   │   │   for j=1 to No_of_fireflies do
15  │   │   │   │   if Ii>Ij then
16  │   │   │   │   │   Compute the attractiveness According to Eq. 8;
17  │   │   │   │   │   Move firefly i from position xᵢ toward position xⱼ;
18  │   │   │   │   │   According to Eqs. 5 and 6;
19  │   │   │   │   end
20  │   │   │   end
21  │   │   │   if rand < mutation_rate then
22  │   │   │   │   Mutate(xᵢ)
23  │   │   │   end
24  │   │   end
25  │   │   Evaluate the current firefly position xᵢ;
26  │   │   Find the current best position x* among all fireflies;
27  │   │   t=t+1 ;                              /* t is the iteration number   */
28  │   end
29  │   Output the best tour x* from all fireflies in the population;
30  end
```

---

4   Experimental Results

Two different experiments were conducted to evaluate the proposed algorithm, HDFA, using different symmetric TSP standard instances, with different length obtained from TSPLIB (http://comopt.ifi.uni-heidelberg.de/software/TSPLIB95/). All of them were run over Intel core-i5 machine. Experimental results for each tested TSP instance were compared with the optimal values in TSPLIB as well as in the literature according to the best, average, percentage deviation of best solution and the percentage deviation of the average solution. The results for each instance were collected after running both experiments ten times. HDFA was initialized with a population of 50 fireflies, 100 maximum number of iterations, 1.0 crossover rate and 0.5 mutation rate.

In the first experiment, three firefly variants with different strategies were developed to evaluate the proposed firefly algorithm. The first one is called FA-opt in which the standard discrete firefly algorithm was hybridized with 2-opt and 3-opt, the second one is called FA-opt-xover in which the crossover was added to FA-opt, and the third one is called by HDFA in which the mutation was added to FA-opt-xover. These variants were evaluated and compared using 12 symmetric TSP instances.

With respect to the best solution, it can be seen from Table 1 that both FA-opt-xover and FA-opt attained similar results to the optimal solution in the following instances (eil51, berlin52, St70, Eil76, kroA100, kroB100, kroC100 and kroe100). However, FA-opt-xover outperformed FA-opt in three instances (pr136, krob200 and pr439). In terms of the average solution, FA-opt-xover outperformed FA-opt in five of the tested instances and gained identical results with three instances. The reason behind that is referred to the adoption of crossover operator which exploits the detected promising solutions to speed up the learning capability of the algorithm. Compared to FA-opt-xover with respect to the best solution, HDFA achieved as the optimal solution for all instances whereas FA-opt-xover failed to obtain such solutions in three instances (pr136, pr299 and pr439). In terms of finding the average solution over the ten runs, both algorithms, FA-opt-xover and HDFA, were identical only in one instance (berlin52). HDFA significantly achieved better solutions than FA-opt-xover for the remaining 11 instances, in ten of which HDFA gained the optimal solution. This superiority of HDFA may be because the introduction of the mutation operator maintains the diversity of algorithm's population, explores new possible regions of the search space and prevents the algorithm from being trapped into local optima.

The performance of HDFA is significantly better than the other variants, FA-opt and FA-opt-xover. It was superior in terms of solution quality with a percentage deviation of best solution PD_{best} of 0.000%, 0.103% and 0.071% for HDFA, FA-opt and FA-opt-xover respectively. In a similar vein, HDFA is superior in terms of solution quality with a percentage deviation of the average solution PD_{avg} of 0.003%, 0.338% and 0.249% respectively.

Table 1 Computational results of the proposed discrete FA variants: FA-opt, FA-opt-xover and HDFA. The best results are given in bold

Instance	Opt. Sol.	FA-opt				FA-opt-xover				HDFA			
		Best	Avg.	PD_best	PD_avg	Best	Avg.	PD_best	PD_avg	Best	Avg.	PD_best	PD_avg
eil51	426	426	426.5	0.00	0.12	426	426.2	0.00	0.05	426	**426**	0	**0**
berlin52	7542	7542	7542	0.00	0.00	7542	7542	0.00	0.00	7542	**7542**	0	**0**
st70	675	675	675.5	0.00	0.07	675	675.6	0.00	0.09	675	**675**	0	**0**
eil76	538	538	540.2	0.00	0.41	538	539.5	0.00	0.28	538	**538**	0	**0**
kroA100	21282	21282	21290	0.00	0.04	21282	21290	0.00	0.04	21282	**21282**	0	**0**
kroB100	22141	22141	22168	0.00	0.12	22141	22172	0.00	0.14	22141	**22141**	0	**0**
kroC100	20749	20749	20749	0.00	0.00	20749	20749	0.00	0.00	20749	**20749**	0	**0**
kroe100	22068	22068	22079	0.00	0.05	22068	22078	0.00	0.04	22068	**22068**	0	**0**
pr136	96772	96887	97122	0.12	0.36	96867	97135	0.10	0.38	**96772**	**96772**	**0**	**0**
krob200	29437	29554	29702	0.40	0.90	29437	29527	0.00	0.30	**29437**	**29437**	**0**	**0**
pr299	48191	48332	48599	0.29	0.85	48388	48640	0.41	0.93	**48191**	**48199.8**	0	**0.02**
pr439	107217	107672	108446	0.42	1.15	107584	108008	0.34	0.74	**107217**	**107239**	0	**0.02**

Table 2 Computational results of HDFA in comparison with DBA and DFA. The best results are shown in bold.

Instance	Opt. Solution	HDFA				DBA				DFA			
		Best	Avg.	PD_best	PD_avg	Best	Avg.	PD_best	PD_avg	Best	Avg.	PD_best	PD_avg
eil51	426	426	426	0	0	426	426	0	0	426	428.1	0	0.49
berlin52	7542	7542	7542	0	0	7542	7542	0	0	7542	7542	0	0.00
st70	675	675	675	0	0	675	675	0	0	675	679.1	0	0.60
eil76	538	538	**538**	**0**	**0**	538	538.76	0	0.14	543	556.8	0.92	3.38
eil101	629	629	**629**	**0**	**0**	629	632.43	0	0.54	643	659	2.18	4.55
kroA100	21282	21282	21282	0	0	21282	21282	0	0	21282	21483.6	0	0.94
kroB100	22141	22141	**22141**	**0**	**0**	22141	22141	0	0	22183	22604.8	0.19	2.05
kroC100	20749	20749	**20749**	**0**	**0**	20749	20753.36	0	0.02	20756	21096.3	0.03	1.65
krod100	21294	21294	**21294**	**0**	**0**	21294	21303.5	0	0.04	21408	21683.8	0.53	1.80
kroe100	22068	22068	**22068**	**0**	**0**	22068	22080.76	0	0.06	22079	22413	0.05	1.54
pr107	44303	44303	**44303**	**0**	**0**	44303	44360.8	0	0.13	44303	44790.4	0	1.09
pr124	59030	59030	**59030**	**0**	**0**	59030	59037.66	0	0.01	59030	59404.3	0	0.63
pr136	96772	96772	**96772**	**0**	**0**	96772	96995	0	0.23	97716	99683.7	0.97	2.92
pr144	58537	58537	58537	0	0	58537	58537	0	0	58546	58993.3	0.02	0.77
pr152	73682	73682	**73682**	**0**	**0**	73682	73759.06	0	0.10	74033	74934.3	0.47	1.67
pr264	49135	49135	**49135**	**0**	**0**	49135	49167.9	0	0.07	50491	51837	2.69	5.21
pr299	48191	48191	**48199.8**	**0**	**0.02**	48191	48311.7	0	0.25	48579	49839.7	0.80	3.31
pr439	107217	**107217**	**107239**	**0**	**0.02**	107291	107683.3	0.07	0.43	111967	115558.2	4.24	7.22

In the second experiment, HDFA, was compared with a well-known discrete bat algorithms (DBA) [20] and discrete firefly algorithm DFA [21] using 18 symmetric TSP instances. Table 2 shows the comparison of HDFA to DBA and DFA with respect to the best and average solution. It could be seen that HDFA outperformed DBA and DFA in terms of the ability of reaching the optimum solution in all instances with respect to the best found solution. In terms of average solution, HDFA reached the optimum solution in most of the instances (eil51, berlin52, st70, eil76, eil101, kroA100, kroB100, kroC100, krod100, kroe100, pr107, pr124, pr136, pr144, pr152 and pr264). For the remaining two instances, pr299 and pr439, HDFA was able to obtain close to optimum solution.

The performance of HDFA is competitive when compared to other existing metaheuristic algorithms such as DBA and DFA. HDFA is superior in terms of solution quality with a percentage deviation of best solution PD_{best} of 0.000%, 0.004% and 0.727% for HDFA, DBA and DFA respectively. In the same vein, HDFA is superior in terms of solution quality with a percentage deviation of the average solution PD_{avg} of 0.002%, 0.113% and 2.212% respectively.

5 Conclusion

In this paper, a new variant of FA called hybrid discrete firefly algorithm, HDFA, was proposed to solve TSP. In HDFA the local search procedures, 2-opt and 3-opt, were used to speed up the convergence, the crossover operator was used to help FA to exploit the current search region efficiently and mutation operator was used to increase the diversity of the fireflies in the population. The experimental study proved that HDFA outperformed the performance of both DBA and DFA in terms of finding the optimal/sub-optimal solution(s) in the most of tested TSP instances. Overall, HDFA is a promising one and has demonstrated a great performance. Results emphasis strongly that HDFA can be used to solve large-scale TSP instances and, consequently, solve more complex optimization problems in the real world applications.

References

1. R. Matai, S.P. Singh, M.L. Mittal, Traveling Salesman Problem, Theory and Applications pp. 1–24 (2010)
2. C. Chauhan, R. Gupta, K. Pathak, International Journal of Computer Applications 52(4) (2012)
3. C.H. Papadimitriou, *Computational complexity* (John Wiley and Sons Ltd., 2003)
4. X.S. Yang, Scholarpedia 6(8), 11472 (2011)
5. I. Fister, X.S. Yang, J. Brest, Swarm and Evolutionary Computation 13, 34 (2013)
6. I. Fister, X.S. Yang, D. Fister, I. Fister Jr, in *Cuckoo Search and Firefly Algorithm* (Springer, 2014), pp. 347–360
7. B. Rampriya, K. Mahadevan, S. Kannan, in *Communication Control and Computing Technologies (ICCCCT), 2010 IEEE International Conference on* (IEEE, 2010), pp. 389–393

8. S.M. Farahani, A.A. Abshouri, B. Nasiri, M. Meybodi, International Journal of Artificial Intelligence **8**(12), 97 (2012)
9. A. Abdullah, S. Deris, M.S. Mohamad, S.Z.M. Hashim, in *Distributed Computing and Artificial Intelligence* (Springer, 2012), pp. 673–680
10. A. Rajini, V.K. David, Int. J. Comput. Appl **30**(6), 10 (2011)
11. X.S. Yang, S. Deb, in *Nature Inspired Cooperative Strategies for Optimization (NICSO 2010)* (Springer, 2010), pp. 101–111
12. G.K. Jati, et al., *Evolutionary discrete firefly algorithm for travelling salesman problem* (Springer, 2011)
13. M. SARAEI, R. ANALOUEI, P. MANSOURI, Cumhuriyet Science Journal **36**(6), 267 (2015)
14. L. Zhou, L. Ding, X. Qiang, in *Bio-Inspired Computing-Theories and Applications* (Springer, 2014), pp. 648–653
15. L. Zhou, L. Ding, X. Qiang, Y. Luo, Journal of Computational and Theoretical Nanoscience **12**(7), 1184 (2015)
16. X.S. Yang, *Nature-inspired metaheuristic algorithms* (Luniver press, 2010)
17. X.S. Yang, *Engineering optimization: an introduction with metaheuristic applications* (John Wiley & Sons, 2010)
18. E. Osaba, X.S. Yang, F. Diaz, P. Lopez-Garcia, R. Carballedo, Engineering Applications of Artificial Intelligence **48**, 59 (2016)
19. P. Thakur, A.J. Singh, International Journal of Advanced Research in Computer Science and Software Engineering **4**(3) (2014)
20. Y. Saji, M.E. Riffi, Neural Computing and Applications pp. 1–14 (2015)
21. G. Jati, R. Manurung, Discrete firefly algorithm for traveling salesman problem: A new movement scheme. Swarm Intelligence and Bio-Inspired Computation: Theory and Applications pp. 295–312

Empowering MOOCs Through Course Certifying Agency Framework

Yeong-tae Song, Yuanqiong Wang and Yongik Yoon

Abstract The arrival of Massive Open Online Course (MOOC) was hailed by many learners around the globe. More universities are willing to offer their top notch professor's courses as MOOC. However, when utilizing the knowledge from MOOCs, learners need to go through a number of hurdles—getting course completion certification (MS Global Learning Consortum: An example of LIP accessibility information, IMS Global Learning Consortium Inc., 2001, [10]) from various providers, get recognition of the knowledge level of a MOOC in the related domain, and make it searchable for various purposes. In this paper, we propose a MOOC course certifying agency framework, which merges learners' profiles from various MOOC providers so consolidated profiles are available in one place. Standards such as IMS LIP and Dublin Core (Feigenbaum and Prud'Hommeaux in SPARQL by example: a tutorial, Cambridge Semantics, 2011, [5]) are adopted and expanded to describe relate MOOC course profiles, learner profiles, learning goals, and related skill sets. It enables matching of qualified learner profiles for a job position and/or to identify a set of related MOOC course profiles for some learning goal. The potential employers look for a matching skill set from converged learner profiles through the agency. Each skill for the position goes through a mapping procedure with a corresponding MOOC course profile. After mapping skills to corresponding MOOCs, the framework searches for the converged profiles. The result is the list of learners who match or almost match a given job description.

Keywords e-Learning · Massive open online course · Learner profile · Course profile · Ontology · Certifying MOOC · Semantic search

Y.-t Song (✉) · Y. Wang
Department of Computer and Information Sciences, Towson University, Towson, USA
e-mail: ysong@towson.edu

Y. Wang
e-mail: ywang@towson.edu

Y. Yoon
Department of Multimedia Science, Sookmyung University, Seoul, Korea
e-mail: yiyoon@sookmyung.ac.kr

© Springer International Publishing AG 2017
R. Lee (ed.), *Applied Computing and Information Technology*,
Studies in Computational Intelligence 695,
DOI 10.1007/978-3-319-51472-7_13

1   Introduction

The introduction of Massive Open Online Course (MOOC) enables vast amount of openly-accessible knowledge sets from many traditional universities. Since its introduction, more than 4200 openly-accessible and college-level courses (from 500 + Universities) have emerged from course providers such as Coursera, Udacity, and edX [18].

MOOCs are beginning to become highly visible and endorsed by many major top-tier universities. The American Council on Education (ACE), an organization that advises college presidents on policy, has gone so far as to endorse five MOOCs from Coursera for credit and is currently reviewing more from Udacity [1, 32]. No longer are MOOCs plagued by the stigma that they no longer offer the quality of education provided by typical university classroom settings.

However, despite the increase in course quality now being seen in MOOCs, these courses offer knowledge in a vast variety of formats and learning pedagogies [3]. Learning material pertinent to a learner's professional development often resides in a variety of locations and is highly disorganized. With learners no longer engaging in traditional learning pedagogies and the weight of accomplishments in the e-learning community no longer easily discerned, several problems have emerged that our approach aims to solve:

- Learners desire to retrieve available courses relevant to their own learning goals potentially in a proper sequence to best achieve their learning goals from a distributed set of MOOC providers and their courses
- There is no way currently to track learners' accumulation of knowledge from different MOOC providers under a single authoritative agency who will facilitate management of this information
- There is lack of agreement among MOOC providers on the approaches or standards to facilitate managing this information

Udacity, Coursera, and edX now claim over 24 million students by EdSurgeNews   (https://www.edsurge.com/news/2015-09-08-udacity-coursera-and-edx-now-claim-over-24-million-students). Without solving the above problems, the knowledge to be gained from completion of these courses represents viable skill sets that may go completely unrecognized once these learners enter the job market. Therefore, having access to a knowledge base of their skill sets could benefit both employers and the students themselves.

Ideally, a MOOC learner should be able to retrieve courses from a variety of MOOC providers where they present a certifiable proof upon successful completion of the course so that the knowledge gained from the MOOC course can be mapped to some kind of skill set that can be recognized by a potential employer or any other entity of interest.

Our approach aims to build a framework that will allow for learners to easily retrieve courses relevant to their specific learning goals in a proper sequence that best suits their learning preferences. The information about successfully completed courses from all MOOC providers will be tracked by an over-arching authority and be searchable by employers looking for employees to fill available job descriptions that require the skill sets may be satisfied by the completed courses.

The rest of the paper will discuss the components of this framework listed as follows:

Section 2 discusses related work and applicable standards for learner profile and course profile, Sect. 3. illustrates learner profile model, Sect. 4 illustrates course profile model, Sect. 5 describes all the constituent components in the proposed model, Sect. 6 briefly describes typical scenarios, and finally Sect. 7 provides conclusions.

2 Relevant Works

MOOC is an educational delivery method that is gaining acceptance in academic circles as an alternative to the traditional instructor led, classroom delivery method. These online courses typically involve videos of lectures combined with interactive assessments [30] while encouraging student collaboration and use of social networking applications. MOOCs combine the connectivity of social networking with the facilitation of an expert in an online, resource rich, environment [15]. The focus on the connectivity of these courses necessitates their scalability and in this case massive truly means massive with enrollments reaching the thousands in a single offering [19]. One important benefit of these courses is that, while they are many times taught by world renowned professors and at highly esteemed institutions, they are typically free of charge. This can be especially beneficial for learners who are lack of funding resources while trying to improve their skill sets to advance their careers. For example, Garrido [8] reported that MOOC has been used for professional development in Colombia, the Philippines, and South Africa. Like any recent innovation, MOOCs are evolving rapidly to suit the needs of both providers and students. As such, MOOC's final form and value has yet to be determined, but the opportunity presented by the MOOC format is attracting a great deal of attention [7].

There are many reasons that the popularity of MOOCs is increasing. With the cost of education rising faster than that of healthcare, MOOC has advantages for students in that they are available anywhere and at no charge [7, 14]. The only prerequisite for the courses is an Internet connection and interest in learning the material. Many students look upon these classes as a way to see if they are interested in a subject without having to pay [22]. If the course is too difficult or not interested in the material anymore, they simply stop attending. The student does not have to worry about a failing grade or an incomplete and is not bound by a financial investment in the course that will be lost by not completing it. It is also important to note that many MOOCs do not list a set of outcomes for qualification of success; students who drop out of the course may have achieved the educational goal they were pursuing by acquiring the desired knowledge [13]. In this case, success or failure of the course is determined by the student's goal and not a syllabus. Thus, a reason for the increasing popularity of the MOOC format is that they present low risk to the student and therefore learning can take place at even a modest interest level [4]. MOOCs are also taught in small chunked lessons that include information and assessments. This chunking of information creates a fast turnaround time between learning a concept and performing the assessment activity [19]. Classroom based courses often feature an hour long lecture and the students are sent home to practice the skills they learned. With the MOOC format the professor typically lectures for a short time then the students perform an assessment, followed by another short lecture and assessment. Assessment activities are designed to encourage learners to be socially active and to pose questions, work through problems, and discuss class topics using social media. This social learning lets the students support and learn from one another.

For the providing institution, they have the ability to attract large numbers of students and crossing national borders. This widespread exposure helps to increase the institution's global awareness and notoriety. Some professors and institutions have recognized that MOOC is an opportunity to take their class worldwide and attract students that would be otherwise unavailable due to geographic or financial factors. Institutions are also using MOOCs as a means of attracting students in the fields of engineering and computer science to their programs [25].

While this growth in popularity is demonstrative of the opportunity that MOOCs may represent, many academics still doubt the effectiveness of the MOOC format [15, 22]. They point out several weaknesses, one of which is the logistics and preparation involved in teaching a class of thousands. Some professors have stated that preparing for a MOOC becomes a fulltime job by itself. Personalized attention is not feasible when classes get that large and even though they spent many hours preparing many professors feel that the students are being short changed by the class [16, 26]. Moreover, students learn in very different ways and not all learning styles can be accommodated in a MOOC [23]. Complex concepts are another problem in that they are difficult to convey even when working in small groups. Some professors who have taught MOOCS have stated that they felt the need to decrease the rigor of the material [28]. The largest recognized evidence that the MOOCs are not as effective as traditional classes is the high percentage of students who do not complete the course [12]. Many professors see this large drop in participation as evidence that MOOCs are not as effective as a traditional classroom based courses [18]. The level of attrition is higher than that of traditional classes and had been shown to be as much as 80%. In one class, 150 thousand students registered and only 20 thousand finished. While 20 thousand students completing an offering of a single class is impressive it is still a decline of roughly 87%. In 2015, a two-year study by Harvard and MIT [9] reported about 57% of the participants stated intention to obtain a certificate from MOOC class, and among the 43% who was unsure or did not intend to earn a certificate, 8% of them eventually did. In addition, they claimed "*increase and formalize the flow of pedagogical innovations to and from residential courses*" as one of the future directions for improving the retention rate for MOOC.

Most professors recognize MOOCs as another tool that can be used to convey information, while MOOC delivery format still needs refinement [2]. Some identified areas where improvement would be beneficial are assessment and certification. Providers and instructors have moved quickly to try to address these weaknesses. Assessments have been improved and even automated; classes taught as MOOCs have had their curriculum aligned with the classroom version in the hope of making the classes comparable. Recent studies have shown students who successfully complete these MOOC courses have no significant difference in later performance than students who complete traditional courses [24]. In one study, students who completed their introductory computer science course through a MOOC were shown to pass more and fail fewer classes than those who completed the face-to-face class. Certification and accreditation of MOOCs could be considered the next step in their evolution. Although, early MOOCs were non-credit or certification courses, many providers including Udacity and Coursera have recognized the need for a verification of completion of the course and are even offering levels of competency such as "Highest Distinction". Thus, providers are able to certify student achievement. Many professors who either have just finished or are in the process of teaching their first MOOCs have stated in a survey that some MOOCS should be counted as regular classes for credits [27, 28]. In February 2013, the American Council on Education (ACE) recommended that its members provide transfer credit from a few MOOC courses [15]. However, majority of the universities are still not accepting it.

3 Learner Profile Model

Currently there is no standardized learner profile format for MOOC providers so it is difficult to combine the content of the learner information and their completed course information from various MOOC providers. There are two competing standards in industry—IMS Learner Information Profile (LIP) and IEEE Personal And Private Information (PAPI). In our approach, we mainly followed IMS LIP and its XML schema format to represent the converged learner profile and course completion information. In IMS LIP, there are 11 core structures to describe a learner such as identifications, security keys, transcripts, goals, qualifications, certifications and licenses (QCL), activities, interest, competency, relationship, affiliation and accessibility [11, 12]. We have added course related attributes to represent course completion information that should come from various MOOC providers' sites:

- Learner: < Identification (Full Name, Email Address)>, <Learning Goal>, <QCL>, Activity (MOOC Completion Status)
- Courses [*]: <Course title> <Course Category>, <Completion Status>, <Date of completion>, <Completion Status (Type of Certificate such as *highest distinction*)>, <Course content provider>, and <MOOC Provider>.

```
<coursetitle>
  <typename>
    <tysource sourcetype="certificate"/>
    <tyvalue>Computer Science 101:Introduction to Computer Science</tyvalue>
  </typename>
</coursetitle>
<coursecategory>
  <typename>
    <tysource sourcetype="certificate"/>
    <tyvalue>Computer Science</tyvalue>
  </typename>
</coursecategory>
<completionstatus>
  <typename>
    <tysource sourcetype="MOOC Site"/>
    <tyvalue>Completed</tyvalue>
  </typename>
</completionstatus>
<dateofcompletion>
  <typename>
    <tysource sourcetype="certificate"/>
    <tyvalue>Date of Completion</tyvalue>
    <certlevel>Certificate of Accomplishment with Highest Distinction</certlevel>
  </typename>
  <date>April 6, 2012</date>
</dateofcompletion>
```

Fig. 1 A part of converged learner profile in XML format

The main purpose of conforming to industry standard such as IMS LIP for the learner profile is to make the system interoperable with other systems e.g. converging learner information from various MOOC providers' sites. IMS LIP specification also defines a set of packages in XML that can be used to import data into and extract data from IMS compliant learning management systems. A partial screenshot of the XML learner profile is shown in the Fig. 1.

4 Course Profile Model

In our proposed approach, we have followed an industry metadata standard - Dublin Core (DC) to represent the course content elements for the course profile. The main purpose of using Dublin Core is to create simple descriptive records for all the MOOC courses. DC elements describe the resources in the networked environment in an effective and interoperable way. It has fifteen "core" elements as shown in Fig. 2 [5]. We have used the following elements to express the course profile contents: title, description, creator, creator organization, type, audience, publisher, identifier, format, typical learning time, workload, category, difficulty level, language and cost. The only added element is 'cost' that we used to describe the price value of the courses. Figure 3 shows our Course Profile in XML format.

Fig. 2 The Dublin Core

```
<?xml version="1.0"?>
<metadata
  xmlns:csmd="http://www.imsglobal.org/profile/cc/ccv1p2/ccv1p2_imscsmd_v1p0.xsd"
  xmlns:xsi="http://www.w3.org/2001/XMLSchema-instance"
  xmlns:dc="http://dublincore.org/documents/dcs/">
  <dc:title>Introduction to Astronomy</dc:title>
  <dc:description>An introduction to astronomy through a broad survey</dc:description>
  <dc:creator>Ronen Plesser</dc:creator>
  <dc:CreatorOrganization>Duke University</dc:CreatorOrganization>
  <dc:type>Lecture Videos </dc:type >
  <dc:audience>Unspecified</dc:audience>
  <dc:publisher>Coursera</dc:publisher>
  <dc:identifier>
    https://www.coursera.org/#course/introastro</dc:identifier>
    <dc:format>text/html</dc:format>
    <ims:typicallearningtime>
      <ims:workload>9 weeks</ims:workload>
    </ims:typicallearningtime>
    <dc:category>Physical & Earth Sciences</dc:category>
    <dc:DifficultyLevel>None</dc:DifficultyLevel>
    <dc:language>English</dc:language>
    <dc:CostCurrency="USD">None</dc:CostCurrency>
</metadata>
```

Fig. 3 Course profile in Dublin Core format

5 Proposed Framework Components

The Course Certifying Agency (CCA) framework consists of various software module components and persistent data storage. Our persistent data storage options utilizes XML technology. The course profiling module continuously monitors newly available MOOCs and populates course profile attributes in an XML data file. *Merging Learner profile* module will update a learner profile when new course completion information for the learner is available from a MOOC provider.

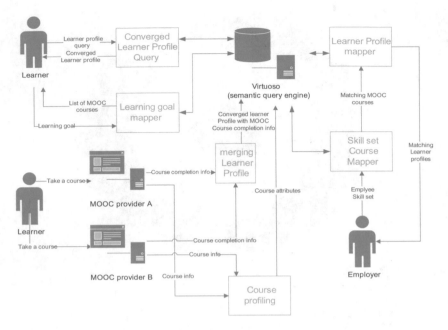

Fig. 4 Course certifying agency architecture

The CCA will gathered data from distributed MOOC providers into both *Merging Learner Profile* module and the *Course Profiling* module. The converged learner profile may be queried by the learners to check their converged profile for their converged MOOC completion information. Learners provide their learning goals to the Learning Goal Mapper to get a list of MOOCs that can help satisfy their learning goals. The course completion information from the learners may be useful for potential employers who are looking for employees with some desired skill set. The desired skill set may be interpreted into a set of relevant MOOC courses by the *Skill set—Course mapper*. The resulting MOOC courses will be utilized in searching a learner's profile that has corresponding MOOC course completion information, which can be done by the *Learner Profile Mapper* as in the Fig. 4.

5.1 Virtuoso RDF Triple Store

The central component used for the distributed learner profile will be an RDF store running an instance of *OpenLink Virtuoso Universal Server* [31]. The Virtuoso RDF Triple Store will house RDF graphs for two ontologies that are direct OWL representations of our Course Profile and Learning Profile models described in Sect. 2 [12]. Using the SPARQL query [6, 26] engine that is built into the OpenLink Virtuoso RDF Triple Store server, we will be able to query the graph for relationships from various MOOC providers to find the courses most relevant to the learner's learning goals or to query learners' profiles for finding best matches with desired skill sets. Additionally, using the built-in OWL inference properties that are possible with Virtuoso, we can expand our queries to result in more related resources in certain ways other than key words. In this case, RDFS and OWL object properties may be included for consideration in the expanded query [16, 17]. For the following examples, URIs will be prefixed per W3C guidelines for the purpose of readability. The two graphs representing our ontologies that model the Learner Profile and Course Profile will be prefixed as "CP:" and "LP:", respectively. "CP:" is a shortened reference name for a URI with a Universal Resource Identifier (URI) such as <http://cp.towson.edu/CourseProfile/CourseProfile.owl>. For the given ontology, properties can be created to describe the relationships between the courses such as CP:isPrerequisteOf and CP:isCorequisiteOf or their inverse properties CP:hasPrerequisite and CP:hasCorequisite, respectively, for a graph prefixed per W3C guidelines as CP [20]. Among many other APIs for RDF data sets, dotnetRDF (http://www.dotnetrdf.org/) was chosen for the manipulation of the proposed data sets. It is an open source.Net API that allows for manipulation of our RDF data sets in a programmatic way. Additionally, the API allows for the parameterized construction of SPARQL queries against the Virtuoso RDF Triple Store [17, 29].

5.2   Course Profiler

A Course Profiler component will perform all duties pertaining to the maintenance of the course information made available from MOOC providers online and converged within the Course Certifying Agency.

5.2.1   Access of MOOC Provider Course Listings via RESTful API

As MOOC providers continue to engage in trending web technologies, it is possible to access course listing for easy integration of our CCA platform via a RESTful API. As an example, the Twitter API documentation provides a RESTful web method "GET statuses/retweets/:id". This method allows users with authentication tokens to substitute an appropriate ID for a given tweet and have up to the first 100 retweets of said tweet returned via JSON format. If such an API were available from MOOC providers, the returned data in JSON format could be packaged via a data manipulation factory class and converted with dotnetRDF into triples nodes representing the subject, predicate, and object of a triples statement and submit it to Virtuoso for insertion into the RDF triple store.

5.2.2 Parsing HTML of MOOC Course Search

With RESTful APIs such as "MOOC LIST (https://www.mooc-list.com/tags/rest-apis)", one can mine course data of MOOC provider sites using *HttpWebRequests* to perform actions against a web site as if it were being performed from a browser. Once this request has been returned to the agent that has spawned the request, the results are handled and parsed.

5.3 Course Profile Mapper

5.3.1 Mapping to Triples Format

The Course Mapper component will interpret the raw data gathered either via JSON or HTML parsing and package the data into RDF triples representing the abbreviated list of Course Profile properties defined in Table 1 below.

The Course Profile structure example in the Table 1 is the example of triple subject-predicate-object statement that we use to define a course listing. In this example a course listing is characterized with a unique identifying integer value that is independent of the MOOC that is originated from our converged profile system. As can be seen, the predicates of these course profile triple statements refer directly to the Dublin Core RDF specification indicated by the prefix "dc:" which is represented in whole by the URI <http://purl.org/dc/elements/1.1/>. This is done to remain in accordance with our standards declared in Sect. 2.

Table 1 Course profile properties and course profile structure (partial example) in RDF triples format

IEEE LOM/DC standards element	MOOC courses elements	*Subject*	*Predicate*	*Object*
Title	Course name with IDs	*<CP:1>*	*<dc:title>*	*<Agile methodology>*
Description	Course overview	*<CP:1>*	*<dc:description>*	*<Introduction to Agile methodology>*
Creator	Instructor	*<CP:1>*	*<dc:creator>*	*<James Bond>*
Creator Organization	Academic Institution	*<CP:1>*	*<dc: creatorOrganization>*	*<Towson university>*
Publisher	MOOC provider name	*<CP:1>*	*<dc:publisher>*	*<Coursera>*

5.3.2 Insert Mapped RDF Triples in Triple Store

dotnetRDF has the ability to designate a set of triples as part of a specific graph. In this instance, as mentioned earlier, the triples mapped using the course mapper will be placed in the CP graph. Figure 5 displays some example code displaying the direct manipulation of triples in a specific graph using the dotnetRDF API and a sample query result is shown in the Fig. 6.

The ontology for Course Profile is defined by owl using Protégé as shown in the Fig. 7 [20, 21].

```
//First connect to a store in this example we use Virtuoso
VirtuosoManager manager = new VirtuosoManager("localhost",
                                    port,
                                    instance,
                                    "user",
                                    "password");

//Construct the Triple we wish to add
Graph g = new Graph();
INode s = g.CreateUriNode(new Uri("http://cp.towson.edu/courseprofile/1"));
INode p = g.CreateUriNode(new Uri("http://purl.org/dc/elements/1.1./#title"));
INode o = g.CreateLiteralNode("Agile Methodology");
Triple t = new Triple(s, p, o);

//Now delete the triple from a graph in the store
if (manager.UpdateSupported)
{
    //UpdateGraph takes enumerables of Triples to add/remove or null to indicate none
    //hence why we create a Triple array to pass in the Triple to be deleted
    manager.UpdateGraph("http://cp.towson.edu/courseprofile/cp.owl", null, new Triple[] { t });
}
else
{
    throw new Exception("Store does not support triple level updates");
}
```

Fig. 5 Code example utilizing dotnetRDF API

```
PREFIX                          CP:
<http://cp.towson.edu/cp/courseprofile.owl>
PREFIX                          LP:
<http://lp.towson.edu/lp/learnerprofile.owl
>

SELECT ?learner_email ?course_id

WHERE  {   ?course_id   CP:hasPreRequisite
CP:Software_Engineering;

        LP:hasLearner

     ?learner_email;


LP:hasTakenCoure CP:272 . }

Endpoint
```

Course_id	Learner_email
720	K1@fakeemail.com
892	K2@fakeemail.com
700	syt@fakeemail.com

Fig. 6 Example SPARQL query for proposed SPARQL endpoint

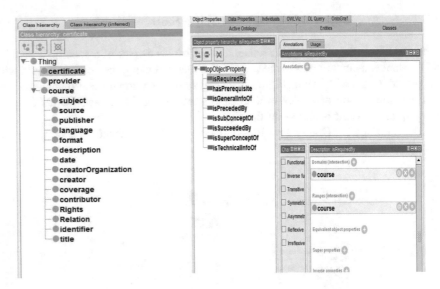

Fig. 7 Course profile owl file in Protégé [21]

5.4 Learner Profiler

Learner Profiler component will perform maintenance regarding the learner's demographics as well as their completed course information for persistent storage from a various MOOC providers. Whenever a learner receives a MOOC completion certificate from a MOOC provider, the profiler updates the Learner Profile so all MOOC completion information may be searched in one place regardless of their providers.

5.5 Learner Profile Mapper

Learner Profile Mapper is a software module that is used when an employer's request comes into find a learner profile that matches with a certain desired skill set for the position. Each skill in the set will go through a mapping process to find a matching MOOC completion record or other related QCL (qualifications, certifications, and licenses) history. The mapping process will have two steps—one for finding match MOOC and the other for searching learner profiles for the identified MOOC or QCL. Any matching over 80% considered relevant for the position. The selected learner profiles need to go through filtering process where location, desired matching level, and any other factors that are imposed by the employer will be utilized. The final result can be exported into serialized XML in the format shown in the Fig. 1 for direct delivery to potential employers for the final selection.

5.6 CCA Component Summary

Utilizing the components of the framework described above, the CCA can perform the following operations:

1. Provide semantic search for all available MOOC provider courses based on learner's learning goal. All MOOCs must be registered in the CCA before use. All object properties need to be set during the registration process for intelligent semantic searches—take advantage of all defined relations among MOOCs.
2. Provide semantic search for employers seeking employees with desired skillsets based on their record of completed MOOC and other QCL. Selected profiles can be delivered to the employers in a serialized XML format.
3. Provide learners the ability to store learner profile information and completed courses from a variety of distributed MOOC providers under one validating, certifying umbrella.
4. Continually update the course profile listings by monitoring a variety of MOOC provider's search sites either via RESTful API techniques or via other crawling techniques.

6 Using CCA

6.1 Setting up Learning Goal

Each learner is encouraged to set up their own learning goals. Each learning goal is interpreted by the CCA and produces a set of MOOCs that help achieve the learning goal. For that, when a course is registered, it is required to set up relationship with other MOOCs such as "superseded by", "isSubconcepOf", "isGeneralinfoOf", or "prerequisite of". Newly identified relationship between MOOCs may be easily added because of the properties—URI—in RDF triple. So when a learning goal is submitted, the course mapper as shown in the Fig. 4 is executed to extract the related keywords for the goal. With resulting keywords, it will execute a SPARQL query with the intention to find all related MOOCs. Once the learner has received a list of MOOCs, s/he will go through filtering for removing unnecessary MOOCs and ordering for the remaining MOOCs for proper sequence.

6.2 Searching for Matching Learner Profile

When needed, employers may request the CCA for the matching learner profile for their advertised vacant position. They may submit a desired skill set(s) for the position(s). The CCA returns the list of learner profiles that match at least 80% of the desired skill set. The requestor may go through the filtering process for additional criteria and finalize their selections.

7   Conclusion

MOOC is becoming a more prevalent trend in postsecondary education and could be future generations' solution for gaining a quality professional education at a low or no cost. MOOCs could function as the bridge between academia and industry allowing learners to customize their skill sets and aptitudes to what prospective employers are searching for. While MOOCs have proved themselves in some scenarios as a valid, credit-worthy endeavor in the eyes of potential employers or even educational councils, there is a need to establish verifiable quality assurance process and also the certification process for MOOC completion so it may become valuable asset for all MOOC learners.

The Certifying Course Agency is an approach that imposes add-on values to MOOC courses so they may be recognized by potential employers as well as by entire learning community. It also facilitates and encourages the standardization of profiles of both courses and learners across distributed MOOC providers. The Certifying Course Agency has proposed current and future methodologies for the aggregation of both Course Listing data and Learner Profile data. With this aggregated data, learners have the potential to search using powerful semantic tools and orient themselves on a path toward their learning goals across various MOOC providers. Going further, the CCA also provides learners and employers a middle-ground to best serve one another both in building better workforces and attaining more fulfilling careers.

References

1. M. Ardis and P. Henderson "Is Software Engineering Ready For MOOCs" *Proc of SIGSOFT Notes*, Cary, NC. ACM 2012 p. 15.
2. J. Calibria, "Connectivist Learning Environments: Massice Open Online Courses", *The 2012 World Congress in Computer Science Computer Engineering and Applied Computing*, Las Vegas, NV. July 2012.
3. D. Clow "MOOCs and the Funnel of Participation" *Proc of the LAK* Leuven, Belgium: 2013 ACM pp. 185–189.
4. Dublin Core Metadata Initiative, "Expressing Qualified Dublin Core in RDF/XML", *Dublin Core Initiative*, 2002 http://dublincore.org/documents/dcq-rdf-xml/.
5. L. Feigenbaum and E. Prud'Hommeaux, "SPARQL by Example: a Tutorial", *Cambridge Semantics*, 2011.
6. J. Flynn, "MOOCs: Disruptive Innovation and the Future of Higher Education" *CEJ Series 3 Vol, 10 No. 1.*

7. Garrido, M., Koepke, L., Anderson, S., Felipe Mena, A., Macapagal, M., & Dalvit, L. (2016). The Advancing MOOCs for Development Initiative: An examination of MOOC usage for professional workforce development outcomes in Colombia, the Philippines, & South Africa. Technology & Social Change Group.
8. Ho, A. D., Chuang, I., Reich, J., Coleman, C., Whitehill, J., Northcutt, C., Williams, J. J., Hansen, J., Lopez, G., & Petersen, R. HarvardX and MITx: Two years of open online courses (HarvardX Working Paper No. 10). 2015.
9. P. Hyman "In the Year of Disruptive Education" Communications of the ACM Dec 2012 Vol. 55 No. 12 pp. 20–22.
10. IMS Global Learning Consortum, "An Example of LIP Accessibility Information", 2001, *IMS Global Learning Consortium Inc.*
11. IMS Global Learning Consortum, "IMS Learner Information Packaging Model Specifications", 2001, *IMS Global Learning Consortium Inc.*
12. R. Kizilcec, C. Piech, and E. Schneider "Deconstructing Disengagement: Analyzing Learner Subpopulations in Massive Open Online Courses" *Proc of the LAK* Leuven, Belgium. 2013 ACM. pp. 170–179.
13. S. Kolowich, "American Council on Education Recommends 5 MOOCs for Credit", The Chronicle of Higher Education, Feb. 2013.
14. S. Kolowich, "The Professors Who Make the MOOCs", *Chronicle of Higher Education*, Nov 2013.
15. Korn, M. Big MOOC coursera moves closer to academic acceptance. Wall Street Journal, February, 7 (2013).
16. G. Lausen, M. Meier, and M. Schmidt, "SPARQLing Constraints for RDF", *Proc of EBDT, ACM*, Nantes, France Mar 2008.
17. F. Manola and E. Miller, "RDF Primer" *W3C Recommendations*, 2004.
18. Dhawal Shah, "By The Numbers: MOOCS in 2015 (https://www.class-central.com/report/moocs-2015-stats/)", December 21, 2015.
19. F. Martin "Will Massive Open Online Courses Change the Way We Teach?" *Communications of the ACM*. Vol 55 No. 8 pp. 26–28 Aug 2012.
20. OML Working Group, "OWL Web ontology Working Guide" *W3C Recommendations* 2004.
21. Protégé, *Stanford University* Accessed on 11/22/2013 Accessed at http://protege.stanford.edu/.
22. A. RIpley, "College is Dead, Long Live College!" *Time* October 18, 2012.
23. Robie, J., Chamberlin, D., Dyck, M., & Snelson, J. XQuery 3.0: An XML query language. Recommendation, W3C. (2014).
24. D. Russel, S. Kelmmer, A. Fox, C. Latulipe, M. Duneier, and E. Losh. "Will Massive Online Open Courses (MOOCs) Change Education" *Proceedings of the CHI* Paris, France ACM Apr. 2013 pp. 2395–2398.
25. M. Sahami, F. Martin, M. Guzdial, and N. Parlente. "The Revolution will be Televised: Perspectives on Massive Open Online Education" *Proc of SIGCSE* Denver, CO ACM Mar. 2013. pp. 457–458.
26. SPARQL Working Group, "SPARQL Query Language for RDF" *W3C Recommendations*, 2013.
27. Taneja, S., Goel, A. MOOC Providers and their Strategies. International Journal of Computer Science and Mobile Computing 3 (5), 222–229 (2014).
28. M. Vardi "Will MOOCs Destroy Academia" *Communication of the ACM*, 2012 Vol. 55 No. 11 p 5.
29. R. Vesse, "dotnetRDF User Guide: Querying with SPARQL" 2013.
30. A. Vihavainen, M. Luukkainen, and J. Kurhila "MOOC as Semester-long Entrance Exam" *Proc of the SIGITE* Orlando, Florida ACM Oct. 2013. pp. 177–182.
31. Virtuoso Server Team, "Openlink Virtuoso Universal Server", Virtuoso, 2009.
32. J. Young "MOOCs Take a Major Step in Qualifying for College Credit" *Chronicle of Higher Education* Nov. 2012.

Author Index

© Springer International Publishing AG 2017
R. Lee (ed.), *Applied Computing and Information Technology*,
Studies in Computational Intelligence 695,
DOI 10.1007/978-3-319-51472-7

Printed in the United States
By Bookmasters